INVENTED BY LAW

Invented by Law

Alexander Graham Bell and the Patent That Changed America

CHRISTOPHER
BEAUCHAMP

Harvard University Press

Cambridge, Massachusetts,
and London, England

2015

Copyright © 2015 by the President and Fellows of Harvard College

ALL RIGHTS RESERVED

Printed in the United States of America

First printing

Library of Congress Cataloging-in-Publication Data

Beauchamp, Christopher, 1977– author.
Invented by law: Alexander Graham Bell and the patent that changed America / Christopher Beauchamp.
  p. cm.
Includes bibliographical references and index.
ISBN 978-0-674-36806-4 (alk. paper)
1. Patent laws and legislation—United States—History. 2. Telecommunication—Patents. 3. Bell, Alexander Graham, 1847–1922 4. Telecommunication—Law and legislation—United States. 5. Patents—United States—History. 6. Inventors—United States—History. I. Title.
KF3116.B43 2014
346.7304'86—dc23   2014012680

Contents

Introduction 1
1. Invention in the Lawyers' World 11
2. Acts of Invention 35
3. The Telephone Cases 58
4. The United States versus Bell 86
5. Atlantic Crossings 109
6. Patent the Earth 130
7. Patents, Firms, and Systems 162
8. Patents and the Networked Nation 185
Conclusion 205

Notes 215
Acknowledgments 261
Index 263

INVENTED BY LAW

Introduction

ON APRIL 8, 1891, dignitaries assembled in Washington, D.C., to mark the hundredth anniversary of a landmark event: the passage of the United States' first patent law.[1] Three days of celebrations were planned, including banquets, receptions, a military parade at the White House, a grand Congress of Inventors and Manufacturers, and a six-hundred-person excursion on the luxury steamer *Excelsior* to Washington's tomb at Mount Vernon. At two o'clock on the first day, attendees gathered for the opening ceremonies at the Lincoln Music Hall, halfway between the White House and the Capitol. The hall thronged with senators, cabinet officials, lawyers, and inventors—"Many Men with Brains," as the *Washington Post* put it, although the newspaper was careful to attribute the brain power to the inventors in the audience rather than to the politicians on the dais.[2] The justices of the U.S. Supreme Court entered as a group, moving in black-robed formation behind the chief justice, to take seats on the stage amid a sustained ovation.[3] Latecomers were still filing in when the president of the United States, Benjamin Harrison, mounted the stage to give the opening address. It had been, the president declared, "a great step in the progress of civilization when the law took notice of property in the fruit of the mind."[4]

After the applause had subsided, a series of speakers rose to tell the story of the patent laws and American invention. One after another, they paid homage to the great inventors and epoch-making technologies: James Watt and Richard Arkwright of England's Industrial Revolution, then Eli Whitney of cotton-gin renown, the steamboat pioneers James Rumsey and John Fitch, the rubber maker Charles Goodyear, and the telegraph inventor Samuel F. B. Morse.[5] The barely unspoken message was that the heirs of these men sat in the audience. Reporters noted the presence of Richard Gatling, creator of the eponymous gun; George Westinghouse, the railroad engineer and electrical inventor; and Emile Berliner, of phonograph fame. Leading technologists of the railroad, telegraph, mining, and machining fields—all shining names of American industry in their time, even if unfamiliar today—rubbed shoulders with figures such as L. E. Waterman, the inventor of the fountain pen, and James L. Plimpton, the father of the modern roller skate.[6]

Another eminent figure, however, proved an "object of more interest in the audience."[7] The celebrity in question was Alexander Graham Bell. His own patent for the speaking telephone, then nearing the end of its seventeen-year term, was the most famous of its day—the reason why, in the words of one speaker, Bell's name was "literally ringing throughout the civilized world."[8] Among the celebrants of the patent system gathered in Washington, Professor Bell occupied a place of honor: he was one of the four vice presidents of the centenary event, a suitable position for the man who had "annihilated space and cuddled the cities of the Republic around a single fireside."[9]

Outside the hall, though, away from the military bands and White House receptions that welcomed the second century of the American patent system, many would have given a darker account of Professor Bell and his patent rights. For years, Bell's claim to the telephone had been assailed in the courts and the press as illegitimate: founded on theft, bribery, and corruption, and propped up only by a "stout wall of fraud and ill-founded judicial decisions."[10] The U.S. government itself had launched a legal attack on the Bell patents, a remarkable and scandalous intervention that had dragged both the inventor and the administration into disrepute. Whatever good name Bell retained was inseparable from his corporate namesake, the American Bell Telephone Company, a nationwide monopolist beset on all sides by rivals and enemies. Mere months after Bell received the accolades of the patent centennial, his own lawyer would wearily observe that "[t]he Bell company has had a monopoly more profitable and more controlling—and more generally hated—than any ever given by any patent."[11]

§

Today, Alexander Graham Bell's invention of the telephone in 1876 is a touchstone of American history. More than any other single event, it serves as a standard example for teaching Americans about the nation's inventive past. Bell routinely ranks among the one hundred "greatest" or "most influential" Americans, whether chosen by historians or by internet polls.[12] His cry of "Mr. Watson—come here—I want to see you," although often misquoted, is one of history's best-known exclamations.[13] Academic and popular-science writers alike have used Bell's telephone to explore the nature of invention.[14] Business writers have looked to Bell's discovery as the origin of a great communications industry.[15] From the History Channel to the halls of academe, there is rarely such consensus about the significance of a technological breakthrough.

To this familiar episode, the chapters that follow bring a new perspective. This book is about law, and its central subject is not Bell's invention, but his patent: the most valuable intellectual property of the nineteenth century and likely the most consequential patent ever granted. The history of Bell's patent rights is, in part, the story of the bitter dispute about who invented the telephone—a famous saga of improbable claimants and scientific quarrelling. The tale of the telephone patents makes a satisfying narrative in its own right, featuring inventors and charlatans, capitalists and politicians, the rise of a business empire, a knife-edge decision at the Supreme Court, and a Gilded Age corruption scandal that reached to the highest levels of government and the judiciary. But it is also a story about how law shapes technology, the economy, and society. Above all, it is a story of patent law, one of the industrial world's most opaque and quietly powerful branches of jurisprudence.

In a sense, the role of law in the history of invention is a problem hiding in plain sight. To any observer of the gaudy 1891 patent centennial, it would have been clear that the patent system was both a prime concern of American industry and a prominent institution of the national government. But even apart from such celebrations, the law kept pressing to the forefront. Patent battles engulfed many of the major technologies of America's industrial age, from the telegraph and electric light to the automobile and the airplane. The historian Daniel Boorstin, surveying these struggles, once observed that "[t]he importance of any new technique in transforming American life could roughly be measured by the quantity of lawyerly energies which it called forth."[16] Historians have not, however, succeeded in explaining why this was so.

This book seeks to illuminate the hidden workings of the patent law in several ways. First, it explains how Alexander Graham Bell came to be anointed the inventor of the telephone through the courts, and why that mattered for the technology that he helped to pioneer. In doing so, it reconstructs the world of nineteenth-century patent law and litigation, seeking both to understand how the law worked and to restore its place in American legal history more generally. At the same time, the story of the telephone patents offers insights into a particular technological and economic moment: the "second industrial revolution" of the late nineteenth and early twentieth centuries. Finally, the view is extended across the Atlantic to place Bell's patent in an international frame. Each of these approaches individually challenges previous historical assumptions. Together, they make an argument that patent law—so often dismissed as an arcane and impenetrable niche of legal practice—has played an active and controversial role in the course of American history.

Placing law at the forefront transforms how we look at the long-contested question of who invented the telephone. Scores of biographies celebrate Bell as the inventor of the device. Conversely, a long line of books question Bell's priority of invention and champion the claims of his rivals. The debunking tradition has flourished in recent years, notably in A. Edward Evenson's closely argued book *The Telephone Patent Conspiracy of 1876* and Seth Shulman's widely noticed *The Telephone Gambit: Chasing Alexander Graham Bell's Secret*.[17]

The popular and scholarly literature about who "really" invented the telephone misses a broader point: that the question itself is a legal artifact. Why, after all, do we still pursue the inventor of the telephone, as opposed to, say, the inventor of the refrigerator or the television? The reason lies in the early patent battles. "Who invented the telephone?" first became a famous question in the 1880s, thanks to the high-stakes and much-publicized litigation that swirled around the technology. Beginning just a few years after Bell's successful telephone experiments in 1876, the Bell Company's attorneys fought a string of cases that sought to bring the entire field of telephone technology under his legal control. Their campaign met with fierce resistance, much of it directed toward disproving Bell's worthiness and advancing the priority of others. On both sides, publicity was a weapon: the glorification of Alexander Graham Bell began as an imperative of litigation; likewise, the promotion of counterclaimants was a necessity for those seeking to challenge the Bell patent monopoly.

More fundamentally, "Who invented the telephone?" was a question defined by law. Legal rules shaped not only the standards of proof but also the terms of inquiry: defining what it *meant* to be a first and true inventor and prescribing the ways that a would-be great inventor needed to describe his achievements in order to gain a patent of maximum breadth. Within these parameters, lawyers prepared the contending positions, marshalled evidence, and argued publicly and bitterly over the origins and nature of the technology.

The lawyers have exerted a powerful grip on historical memory. Today we may take for granted that the telephone originated with a single man, that it consisted essentially of a single invention, and that it represented a sharp technological break with the prior art. But for Bell and his legal representatives these were bold arguments, deliberately and consciously made in pursuit of a patent that would control the telephone business. In the courts of the day, and in the judgment of posterity, those arguments succeeded in spectacular fashion. In that sense, it was the lawyers, as much as anyone else, who invented the telephone.

§

What makes this story matter beyond the telephone case is the window it gives us into the law of the age. The Bell litigation was one of the largest courtroom conflicts of any kind during the nineteenth century. Yet one would hardly know it from reading a standard legal history of the period. This omission is all the more striking since, for decades, one of the central concerns of American legal historians has been to explore the economic effects of the law and the role played by legal institutions in economic development.[18] Patent law involves issues that should be central to these debates. Still, one can search in vain for a discussion of patent law in the classic works on law's relationship to the economy.[19] Patent law has been eclipsed, not only by the great issues of constitution-making, war, race, and slavery, but even by less romantic matters such as insurance and accident law, the law of watercourses, bankruptcy, and regulation.

It is not hard to guess why. Patent practice has long labored under a reputation for inaccessibility, professional specialization, and narrowly fact-specific court rulings. All these factors have functioned to distance patent law from the historical mainstream. Recent years have seen a number of important scholarly inroads into the social, political, and ideological landscape of the patent system.[20] Even so, the historical law of patents remains in many ways unmapped, and its connection to the broader setting of legal and political institutions unclear. This obscurity is unfortunate, since patent law was far from being a legal backwater. Instead it was a ubiquitous feature of rapid technological development, a palpable force in the lived experience of American industrialization, and an integral part of the law.

For a start, it would be hard to overstate the number of major new technologies that were subject to patent litigation in this period. Well-known bids for legal control such as Bell's claim to the telephone, Morse's to the telegraph, Edison's to the electric lamp, and the Wright Brothers' to the airplane are just the beginning. A short list of other highlights would include legal battles over the waterwheel, sewing machine, mechanical reaper, barbed wire, baking soda, fountain pen, typewriter, cash register, phonograph, bicycle, and automobile. These conflicts were not mere intramural disputes among inventors and manufacturers; they were often highly public matters, bound up in the heated politics of monopoly, and regular fodder for scandal, agitation, and congressional intervention.

At the same time, patent practice was a prominent and well-integrated branch of the law. An astonishing proportion of federal cases in the major industrial jurisdictions were patent cases: in some courts, patent matters made up a third or even a half of the decisions published by the court reporters.[21] Throughout much of the nineteenth century, patent business was a staple activity of the elite bar. Senators and cabinet officials routinely argued patent suits in the Supreme Court and lower tribunals. One indicator of the reach and prestige of the field is that no fewer than three members of President Lincoln's cabinet had been involved in high-profile patent matters, as had Lincoln himself.

Patent law is overdue, then, for a move into the historical limelight. To the uninitiated, the story that follows offers a guide through the world of courts and litigation strategy, subjects that have hitherto confused rather than enlightened many students of the history of technology. Patent litigation need no longer be a "black box"—the technologists' term for a process whose inputs and outputs are observed, but whose inner workings are not. To aficionados of law, this narrative gives a new vantage point for considering old questions. Patent law raises many of the same themes that have traditionally structured the field of legal history, including the role of law in economic development, the responsiveness of legal institutions to broader changes in society, and the tension between the law on the books and the law in action. The Bell story suggests that there is no clear, unmediated relationship between the patent system and economic change. That there was a relationship, however, is revealed by the parts that patents came to play in the transformation of the industrial economy.

Looking back from our present age of scientific, large-scale, corporate research and development, Alexander Graham Bell is sometimes seen as an archetype of the independent inventor—the kind of figure for whom the patent system was designed. Since "[t]he prototypical innovation contemplated by the patent law is made by an individual inventor," modern legal scholars have suggested, "Alexander Graham Bell is in many ways the icon of the patent system."[22] To his contemporaries, however, Bell and his patents came to represent something quite different: the growing use of intellectual property by large corporations.

Bell received his patents at a time of rapid industrial change. The late nineteenth century witnessed a wave of new inventions and new business forms that is often described as the beginning of a "second industrial revolution." Like the first industrial revolution of iron, steam, the joint-stock company, and the factory system, the second industrial revolution consisted of linked technological and institutional developments. One was the opening of new industrial sectors, especially in science-based fields: electrical light, power, and communications; motorized transportation; synthetic chemicals; and so on. The other major development was a rapid rise in the scale of corporate organization—crudely put, the rise of "big business"—which gave firms new capabilities to control markets and to manage technology. Thanks to their position at the intersection of these two trends, patents became enormously important to the organization of the new economy.

Gauging the effect of patents on society means looking at how they were exploited in practice, not just how they were litigated in the courts. Accordingly, the story of Bell's patent rights leads us into one of the great stories of American business history: the rise of the American Bell Telephone Company, later and better known as AT&T, which dominated the country's telephone service for a century until its breakup in the 1980s. American Bell was an enterprise constructed around patents; its business strategies, corporate organization, and financial prospects largely depended—at least at first—on legal ownership of key inventions. From about 1880 until the early 1890s, American Bell exercised monopoly control of the telephone industry by laying exclusive claim to the transmission of speech by electricity and forcing all potential competitors from the field. Patent power was placed squarely in the limelight. When Alexander Graham Bell's rights came before the U.S. Supreme Court in 1887, they sustained a "hundred-million-dollar" corporation and made for a trial of unprecedented size, expense, complexity, and controversy.[23]

Without the Bell Company's early dominance, the entire evolution of American communications could have taken a quite different track. This book is the first to explore fully the role of the Bell patent rights in that process: not only by explaining the court judgments that created the monopoly, but also by tracing the role that patents played in forming the emerging telephone industry. An argument for the deep embeddedness of patents in the strategy and structure of a high-technology industry may seem straightforward, but it actually represents a departure in the history of both the telephone and the "rise of big business." Historians have tended to treat patents either as an aspect of the innovation process or as an occasional and awkward weapon of market competition. Organization without patents produced great corporations, they (rightly) point out; patents without organizations never did. But one can accept these observations and still arrive at a fuller appreciation for the multiple uses of patents in the growing corporate economy, as tools of contracting, alliance, capital formation, and technological cooperation. True, patents did not *make* big business big, but they helped to shape the development of the new industrial giants.

Finally, the Bell story would be incomplete without some attention to the world beyond the United States. The second industrial revolution was an age of globalization, characterized by growing international flows of trade, capital, labor, and technology. Inventions moved readily between countries, borne by travel, trade, and the beginnings of multinational industrial enterprise. National experiences were shaped by continuous exchanges of people, technology, and ideas. In the case of the telephone, these transnational elements are inescapable. Not only did the American invention cross borders within its first few years of existence, but so did its inventors and commercial pioneers, its business models, and its patents. These international connections and transitions are as ripe for study as the national industries they produced, and—my central interest in this work—as much shaped by law.

A global history of the telephone industry is beyond the scope of this work. Instead, the book will focus its overseas sojourning on one country, enough to gain comparative and transnational leverage while respecting the details of the local story. That country is Britain, by far the most lucrative overseas market for the telephone, and the site of a telephone patent struggle dramatic and vicious enough to rival that of the United States. Between the United States and Britain lies a fascinating contrast of legal systems, polities, and cultures, crosscut by the commercial adventures of the patent-holders (in Britain, these were primarily Bell and Thomas Edison) who fought for monopoly. The underlying technology was identical in both countries, yet understood quite differently by their respective courts, which provides a kind of experimental "control" for examining the two national patent systems.

Patent law thus reminds us that, however global and apparently seamless the spread of an industrial revolution, nation-states still played a central role in the process. Identical basic inventions supported different configurations of patent rights (or a lack thereof) from one country to the next, helping to decide which companies would commercialize the technology and under what competitive conditions. These rights depended on national legal systems for construction and enforcement. For all the speed with which its major technologies and corporate actors diffused internationally, the second industrial revolution was significantly influenced by its local legal environments.

For much of the past hundred years, it would have been hard to convince people that patent law was a subject rich in drama. Today, the prospect is less daunting. The massive expansion of patenting and patent litigation in the early twenty-first century has produced a slew of much-decried results: billion-dollar judgments; opportunistic "patent trolls"; costly, unpredictable litigation; and an ever-rising flood of patents— some trivial, some promising enormous riches, and some both. Although it remains possible to explain and extol the social benefits of patents as tools of innovation policy, a gloomier view has become widespread. The institutions of the patent law, once feted for adding "the fuel of *interest* to the *fire* of genius," have instead turned sinister: "Judges, bureaucrats, and lawyers," as one critique of the U.S. patent regime warns, "put innovators at risk."[24]

Neither the defenders nor the detractors of the patent system will find unambiguous confirmation in these pages. The patents described here rewarded invention, protected a disruptive new technology from hostile incumbents, and made possible grand schemes of engineering and corporate construction. They also monopolized an industry, confounded the courts, and placed the fate of a transformative communications technology in the hands of the lawyers. For anyone who seeks a modern lesson in this story, it may well be the simple one: all this has happened before, if not in its precise details, then at least in familiar outlines. Conflicts within and about the patent system are an old thing in America and, comfortingly or otherwise, were probably more wrenching in former times than in our own.

CHAPTER 1

Invention in the Lawyers' World

IN 1836, AN IMPOSING BUILDING began to rise in the center of Washington, D.C. The site, seven blocks east of the White House, had been specially reserved in Pierre L'Enfant's original plan for the city. L'Enfant had proposed a patriotic church, "intended for national purposes, such as public prayer, thanksgiving, funeral orations, &c., and assigned to the special use of no particular Sect or denomination, but equally open to all." This pantheon-like structure would also house monuments to the heroes of the Revolutionary War and to "such others as may hereafter be decreed by the voice of a grateful Nation."[1] No national church was built, and for decades the land was used as an orchard.[2] Finally, the site was dedicated instead—some thought appropriately—for the construction of the new Patent Office.

Within a few years, the office's two-block-wide South Wing and massive Greek Revival–style portico loomed over the neighborhood of low brick and timber houses. The patent-examining staff moved there in 1840, after which the building steadily began to fill with clerks, files, technical drawings, and roomfuls of the miniature wood-and-brass models that patent applicants were required to provide with their applications. Various items of national importance were transferred to the carefully fire-proofed facility. The collection included the original Declaration of Independence, which from 1841 to 1876 hung framed in a third-floor display cabinet, slowly fading to illegibility in the sun.[3] By midcentury, a hundred thousand people were said to visit each year to view the artifacts on display.[4] Construction continued around them. The whole structure would not be complete until after the Civil War, by which time the Patent Office complex was reportedly the largest office building in the country.[5]

If the U.S. government of the nineteenth century was a deceptively low-profile force—"a government out of sight," as one scholar has dubbed it—then the patent system was one of its most visible manifestations.[6] The hulking Patent Office in the capital was only one piece. Elsewhere, the patent system existed wherever a federal court sat, a patent lawyer kept an office, or an inventor could mail a package of papers to Washington along with the $30 or $35 application fee.[7] Hundreds of thousands of Americans applied for a U.S. patent between 1790 and 1900, collectively receiving some 650,000 issued patents. Millions more used patented technologies. Most doubtless did not notice or care about the intellectual property rights attached to their everyday items, despite the "U.S. Patent" markings stamped on their sewing machines, tools, firearms, and railroad cars, and printed on the packaging of their food and drugs.[8] But many did, thanks to the patent wars that flared up around them.

Among those conflicts, the telephone patent fights may have been unusually large and complex, but they were part of a broader phenomenon. Major new technologies were commonly patented in nineteenth-century America. "Great inventors" (a subjective term to be sure, but here meaning those later hailed as such by historians and biographers) were, as a class, closely engaged with the patent system.[9] Men of genius were not the only beneficiaries of the law. Any lucrative new industry was an invitation for opportunistic patent claimants, or the purchasers of patents, to come forward in the hope of seeking windfall gains. As a result, new technologies characteristically experienced a phase of heavy litigation, along with intense trading and speculation in patents, while the various pioneers and followers in the field clarified their respective legal entitlements. Massive campaigns of nationwide litigation were not uncommon: the Bell Company filed around six hundred suits to enforce its basic patent, but that was far from being the most prolific enforcement campaign.

Acts of invention were as varied as their authors. They occurred in cities and in backwaters, produced technologies large and small, and sprang from amateur tinkerers and college-trained scientists. No single milieu captured their diversity—except one. Again and again, inventions were led into the lawyers' world.

§

The United States passed its first patent law in 1790. The law's origins were pragmatic. Among the leaders of the new nation, the development of an independent national industrial capacity was a pressing concern. Figures such as Benjamin Franklin, a longtime organizer of associations for the promotion of the "useful arts," and Alexander Hamilton, an ambitious promoter of industrial ventures as the first secretary of the treasury, privileged the acquisition of European technology and the creation of a domestic manufacturing base.[10] The flood of British imports that followed the end of the Revolutionary War served to warn, as the revolutionary leader Benjamin Rush had pointed out on the eve of independence, that "A People who are entirely dependant [sic] upon the foreigners for food or clothes must always be subject to them."[11] In this view, industrial growth offered a bulwark of economic security in an Atlantic world still disturbed by mercantilism and war.

One tool of industrial promotion familiar from British usage was the patent grant: a royal conferral of exclusive rights granted in return for the invention or introduction of a new technology. Monopoly and monarchical privilege were generally anathema to the recent revolutionaries, but "letters patent" for invention had a more encouraging heritage. They had been pointedly excepted from the ban on royal monopolies that Parliament had forced on the crown in the 1624 Statute of Monopolies.[12] In the seventeenth century, the patent had remained largely a tool of royal policy, but during the eighteenth century the grant had evolved haphazardly into something more utilitarian. English judges had begun to speak of a bargain between the inventor and the public, comprising the disclosure of the invention in return for the temporary grant of an exclusive right.[13] As one jurist put it, "The consideration, which the patentee gives for his monopoly, is the benefit the public are to derive from his invention after his patent is expired."[14] As the Industrial Revolution gathered steam, it became easier to associate the system of letters patent with Britain's rapid industrial progress. Several of the states had already issued legislative patents on a case-by-case basis after independence.[15] James Madison, writing to Thomas Jefferson, captured the cautious sense that the patent was a tame monopoly. "Monopolies . . . are justly classed among the greatest nusances [sic] in Government," he admitted. "But is it clear that as encouragements to literary works and ingenious discoveries, they are not too valuable to be wholly renounced?"[16]

Accordingly, the Constitutional Convention at Philadelphia discussed a federal patent power alongside proposals for a national university, prizes for inventors, copyright protection for authors, and other material inducements to the growth of useful knowledge.[17] Most of these provisions fell out of the drafting process, leaving only the patent and copyright powers. The Convention eventually adopted them together as Article I, Section 8, Clause 8 of the Constitution, empowering Congress "To promote the Progress of Science and useful Arts, by securing for limited Times to Authors and Inventors the exclusive Right to their respective Writings and Discoveries."[18] The first Congress, with some prodding from President Washington, implemented the mandate by passing the 1790 Patent Act.[19]

The 1790 law bore little resemblance to a modern, bureaucratic patent system. Applicants for patents submitted their petitions to a cabinet-level board consisting of the secretary of state, the secretary of war, and the attorney general. Any two of these high officers could issue a patent of up to fourteen years' duration to the inventor of a "useful art, manufacture, engine, machine, or device . . . not before known or used," providing that they "deem[ed] the invention sufficiently useful and important."[20] The members of the board exercised their discretion vigorously, scrutinizing each petition and granting fewer than half of the applications they received.[21] But the practice of having patents reviewed by the republic's leading officials soon proved impractical. Thomas Jefferson, who as secretary of state had taken the lead role, was complaining by 1792 of being "oppressed beyond measure" by a task that required him to give "undue & uninformed opinions" on subjects he had no time to master.[22] In addition, Jefferson later noted, the board had collectively "done but little" to organize a body of rules or procedures for patent grants.[23]

After fifty-seven patents and fewer than three years, the 1790 Act was replaced by a much more open system. The Patent Act of 1793 introduced what was essentially a registration regime.[24] No examination took place before issue; applicants simply submitted petitions describing their inventions, swore an oath that they believed themselves to be the true inventor, paid the (then substantial) $30 fee, and received a patent. A handful of State Department clerks handled the processing. From 1802 they were overseen by a superintendent of patents, William Thornton, who became the key administrative figure of the patent system's early years. Thornton was a politically connected polymath: an intimate of Madison and Jefferson, he was an intellectual, a physician, a gentleman architect, an early promoter of steamboats, and had served as one of the commissioners responsible for planning the new federal city at Washington.[25] During a quarter century at the Patent Office, Thornton worked fervently to further the interests of worthy inventors, for whose sole benefit he believed the law had been designed. Many of his procedural innovations were designed to bring inventors greater security in their rights. So were his occasional subversions. He refused, for example, to provide outsiders with copies of any patents that were still in force, despite a widespread understanding that early disclosure was part of the patent bargain.[26]

Embryonic though it was, the early patent system quickly attracted conflict. The first major fight took place over the invention of the steamboat, which was contested between John Fitch, James Rumsey, and a handful of other claimants.[27] Fitch and Rumsey had experimented with steam-powered navigation in the 1780s, while simultaneously jockeying for government support. Both had sought and received a number of state patents.[28] Both also approached the national government in the years before the adoption of the federal patent law. Rumsey was the first, petitioning the Continental Congress and gaining the somewhat distracted favor of George Washington.[29] Fitch was the more ambitious, hauling his boat to Philadelphia to display to the members of the Constitutional Convention in 1787, and asking Congress for a broad patent over "the principle of applying the power of steam to the purposes of navigation."[30]

The eventual fate of these claims reflected the fumbling state of the new federal patent law. After the passage of the 1790 Act, Fitch and Rumsey made new applications and secured a hearing before the patent board. Jefferson and his colleagues, however, lacking any clear criteria for determining priority, declined to prefer any one patent over the other, leaving neither claimant satisfied.[31] Without the exclusivity they had hoped for, Fitch and Rumsey drifted rudderless: Rumsey roamed in Europe, seeking capital; Fitch retreated embittered and abandoned his invention. Steamboat technology stagnated and did not revive until Robert Fulton entered the field in the 1800s.[32]

Two other notable disputes also highlighted the confusion of the early patent law. In Georgia in 1793, transplanted New Englander Eli Whitney built an improved form of cotton gin—not the first cotton gin, but one that worked on a new principle and offered greater productivity than the gins then available.[33] He then returned to Connecticut to manufacture the machine for wider use. He obtained a federal patent early in 1794 and began a years-long process of trying to enforce his patent rights in the South, even as new gins (not all on Whitney's model) started to revolutionize cotton agriculture.[34] A series of lawsuits filed in Georgia yielded only frustration at first.[35] Whitney's partner Phineas Miller reported that "[s]urreptitious gins are erected in every part of the country; and the jurymen at Augusta have come to an understanding among themselves, that they will never give a verdict in our favor, let the merits of the case be as they may."[36]

Worse still, it became apparent that the language of the patent statute prevented meaningful enforcement. The Act of 1793 forbade any person to "make, devise and use, or sell" the protected invention, on pain of damages.[37] The judge in Whitney's first trial held that an infringer must thus both make *and* use the device to infringe, whereas the accused parties had merely purchased and used devices constructed by others.[38] Only in 1800 did Congress amend the statute to render mere use an independent ground for infringement.[39] Whitney prevailed in a few suits in 1807, but by then his patent was just a year from expiring. In the end, such gains as Whitney and Miller would make from the patent came largely from the negotiated sale of rights to state governments.[40]

While Whitney struggled in the South, another ambitious patentee sought to establish his rights in the Mid-Atlantic states. Oliver Evans was a prodigious Delaware-born inventor who made significant advances in both flour-milling and steam engine technology. He too held a number of patents granted by state legislatures in the 1780s, and with the advent of the federal patent law sought national protection as well.[41] Evans received the third U.S. patent issued under the 1790 Act, covering a mechanized system for flour manufacturing. Although the patent was vague about the fact, most of the component machines of Evans's system were old; his contribution consisted of the combination arrangement itself, and also of a revolving rake machine called the "hopperboy." At first, his commercial success was limited. Evans engaged licensing agents in six states, but millers were slow to adopt the labor-saving machines.[42]

What transformed Evans's patent into the most notorious of the early republic was an extension of its term. The grant expired in 1805, leaving some holdover litigation for earlier infringement. In 1807, a federal circuit court in Philadelphia found the patent invalid due to an incomplete disclosure in the final patent document, apparently caused by the issuing officials' failure to include details given in Evans's application.[43] Evans quickly mobilized President Jefferson to write that a bureaucratic failure to comply with the statute's requirements "cannot invalidate the inventor's right, who has been guilty of no fault," and persuaded Secretary of State James Madison to inform Congress that the court's interpretation "would admit the invalidity of all the patents issued in the same form" under the 1790 Act.[44] This statutory mess coming to light over an expired patent proved an enormous boon to Evans. In 1808 a sympathetic Congress granted an "Act for the Relief of Oliver Evans," reissuing his patent for an entire new term of fourteen years.[45]

Evans then began a second campaign of enforcement much more controversial and aggressive than his first. His agents spread across multiple states to demand license fees from millers—higher fees than before, now that the technology had spread—and to institute litigation where necessary. The number of suits is unknown, but was evidently substantial: Evans even printed up a standard form complaint, with a blank to fill in the name and details of the defendant.[46] The volume of litigation was certainly enough to make a mark on the case law of the early patent system. The record of early nineteenth-century published judicial decisions is generally sparse, but it includes eleven cases on the Evans patent, seven of them trial court decisions (all from Maryland, Virginia, and Pennsylvania) and four from the Supreme Court—four, in fact, of the high court's first five decisions in patent cases.[47] In total, only 43 patent cases were published in the law reports in the first two decades of the nineteenth century, of which Evans cases made up more than a fifth.[48] How much any of this benefited Oliver Evans is unclear. The inventor won multiple judgments in the thousands of dollars and fought off attempts in Congress to overturn his extended grant.[49] He also inveighed constantly against the shortcomings of the patent system and experienced maddening reverses in court, including one in 1809 that drove him to burn piles of his inventive notes in a rage.[50] Perhaps fortunately, he did not live to see the Supreme Court invalidate the flour-milling patent in 1822, on the grounds that he had not precisely specified the part of his machinery that was new.[51]

§

Asserting rights over such strategic technologies as steam navigation, cotton ginning, and mechanized milling had proved difficult amid the growing pains of the patent law. Even so, the travails of high-profile patentees did not stop a steadily increasing number of patentees from registering their inventions under the 1793 Act. Nearly a thousand patents were issued during the 1800s; nearly two thousand in the 1810s.[52] More than anything, trends in patenting traced patterns of economic growth: rising during a period of prosperity in the first decade of the century, then falling back during the War of 1812 and the contraction in commerce that followed; growing rapidly again in the 1820s and early 1830s, before dipping with the financial panic and depression of 1837–1842.[53] Similarly, the geographic distribution of patent applicants reflected the importance of access to markets. Cities had the greatest concentration of patentees, despite harboring a small minority of the population. Patenting in nonurban areas clustered near waterways, which in this period provided both the principal source of mechanical power and the principal means of rapid transportation.[54]

What all this suggests is that patenting activity grew less from specific developments in technology than from broad-based economic factors: market size, demand for goods, flows of information and of capital. Patented invention reflected the widespread release of creative energies. Most patents in the first half of the nineteenth century were issued to inventors who received only one or two patents during their lifetimes. The bar of technical expertise to obtain a patent was not particularly high. Even among urban patentees, skilled engineers and machinists were a minority; more recipients of protection were drawn from the merchant or general artisan classes.[55] Centers of organized industrial patenting did exist: in the nascent New England textile industry, for instance, the heavyweight Boston Manufacturing Company was systematically seeking grants as early as the 1810s.[56] But as a general matter, the patent system was an open one—"democratic," as historians have noted, in its scale and accessibility.[57]

Over time, the democracy of invention began to appear disorderly. The lack of any prior examination for patents under the 1793 Act meant that any issued patent carried with it considerable uncertainty. Compliance with the threshold requirements of the statute (such as the invention being "new and useful" and "not known or used before the application") had to be resolved through litigation, as did conflicting claims among inventors. As a result, complaints about the low quality of patents were widespread. In 1809, Superintendent of Patents William Thornton complained that "many of the patents are useless, except to give work to the lawyers, & others so useless in construction as to be . . . merely intended for sale."[58] Cautionary tales circulated of "frivolous, absurd, and fraudulent" patents asserted in nuisance suits.[59]

In 1826, Judge William Van Ness of the federal district court in New York, deciding a case involving a carpet patent, summed up contemporary criticism in a lengthy broadside. He began by observing that "[m]ore than three thousand patents have been granted since the year 1790," many of them for "the same or similar objects. . . . Eighty are for improvements on the steam engine and on steam boats; more than a hundred for different modes of manufacturing nails; from sixty to seventy for washing machines; from forty to fifty for threshing machines; sixty for pumps; fifty for churns; and a still greater number for stoves." He continued,

> The very great and very alarming facility with which patents are procured is producing evils of great magnitude. It encourages the flagitious peculations of imposters, and the arrogant pretensions of vain and fraudulent projectors.... Amidst this strife and collision, the community suffers under the most diversified extortions. Exactions and frauds, in all the forms which rapacity can suggest, are daily imposed and practised under the pretence of some legal sanction.... Impositions of this sort, are of common occurrence, and will continue to multiply while the door to imposture is left open and unguarded.[60]

Van Ness was no avowed enemy of the patent law or of meritorious inventors. With the same breath, he extolled "the great principle and design of the act" and lamented the fate of "Whitney [and] Evans ... sunk under vexation and the pressure of litigation." He felt compelled, however, to join those calling for "[s]ome mode ... of examining into the novelty and utility of alleged inventions, before patents are issued to the applicants."[61]

Following years of complaint about the defects of the law, Congress eventually moved to replace the registration system in 1836, installing in its place a Patent Office with a professional staff empowered to examine applications before issue.[62] By providing an official check on the validity of grants and by creating an agency that would formalize, coordinate, and promote patent practices and the dispersal of patent information, the act greatly increased the value and efficacy of the system as a whole. With the 1836 Patent Act, the United States had created the first truly modern patent system. The transition between regimes received dramatic punctuation on December 15 of that year, when the building temporarily housing the Patent Office and the General Post Office burned spectacularly to the ground. Of the models and files gathered there in the first forty-six years of the patent law, not a stick or page survived.[63]

§

Under the auspices of the 1836 Act, patents gradually took up an increasingly prominent position in American life. More were sought, and more granted. After an initial dip caused by the 1836 tightening of standards and then the economic downturn following the Panic of 1837, the patent system resumed rapid growth in the mid-1840s.[64] The Patent Office in 1849 received nearly 2,000 applications and issued nearly 1,000 patents, both more than double the numbers of a decade earlier. During the course of the 1850s those annual numbers more than tripled, to more than 6,000 applications filed in 1859 and more than 4,000 patents granted.[65]

Similarly, the institutions of the patent system continued to take shape in the 1840s and 1850s. One of the most important developments was the rise of a cadre of patent professionals, comprising both examiners and patent agents, who collaborated to synchronize the rules of the Patent Office and the needs of inventors. As middlemen, this group performed two related functions. First, they translated between the legal and the technical spheres of patent practice, working with inventors to draft applications and shepherding those applications through the Patent Office examination process.[66] Second, they acted as intermediaries in the market for inventions: promoting and brokering the sale and assignment of rights by inventors, monitoring the landscape of new technology, and evaluating and recommending the purchase of rights by firms.[67]

What really raised the salience of patent law at midcentury, though, were the large-scale campaigns of enforcement launched by a few patent owners. The details of these efforts varied. Some involved technologies from the first wave of American industrialization; others more recent inventions. Most took advantage of one or more of the pro-patentee tools available in patent law in the middle nineteenth century, especially term extension and reissue. Most involved politics in one way or another. Collectively, the weight of litigation they generated shaped the case law and the political economy of patents.

One group of patents to roil the courts after the 1836 Act concerned ubiquitous technologies of American industrialization: water power and woodworking. These patents belonged, in a sense, to an earlier age. They had been issued under the previous legislative regime and dated back to the 1820s and even earlier. Given the standard fourteen-year term of a patent, such patents would ordinarily have expired by the early 1840s at the latest. Their continued prominence in midcentury law rested on extensions of their term—a feature that made them unusual among patents generally, but characteristic of those that generated the most litigation in the 1840s and 1850s. Term extensions could be granted by Congress or, after 1836, by a board consisting of the secretary of state, the solicitor of the treasury, and the commissioner of patents (and after 1848 by the commissioner of patents alone). The theory in either case was to reward a deserving patentee who had "without neglect or fault on his part . . . failed to obtain, from the use and sale of his invention, a reasonable remuneration for the time, ingenuity, and expense bestowed upon the same."[68] Term extension raised the stakes both for and against a patent: it allowed patentees to enforce their rights over a more mature and more widespread (and thus more valuable) technology, and it stimulated legal and political resistance to monopolies that were widely attacked as illegitimate.

One leading example was that of Austin and Zebulon Parker, brothers and millwrights from Ohio, who developed an improved reaction waterwheel in the 1820s and obtained a patent for their invention in 1829.[69] The wheel provided superior power generation, and its use spread among water mills. In 1843, as their term expired, the Parkers acquired a seven-year extension from the Patent Office. This much was unremarkable. What was striking was the scale of their campaign to enforce the patent. Agents of the Parkers spread across the country, demanding license fees of between $10 and $50 per mill. In Lycoming County, Pennsylvania, for example, a correspondent of the journal *Scientific American* noted that Parker's representatives had visited "all, or nearly all, of the saw mills in this section" during 1848.[70] Ohio and Pennsylvania were the most thickly blanketed, but tales came in of agents as far afield as Vermont and New Hampshire.[71] Where mill owners resisted, litigation followed. By 1845, more than a dozen judicial decisions on the Parker patent had been published—at a time when law reporting was still scarce.[72] In 1849, a reporter's note on one case identified more than 200 Parker cases then pending in Ohio alone.[73] The following year, their relative Oliver H. P. Parker, who held the rights for the patent in five states, filed 150 suits in the federal court in Philadelphia.[74]

The Parker patent's mass litigation was not alone. Two other patents controlled important advances in the mechanization of woodworking technology. Thomas Blanchard's turning lathe enabled the shaping of wood into irregular forms such as gun stocks, tool handles, shoe lasts, and hat blocks, reducing to a ten-minute task what might have taken a skilled last-maker hours to complete using hand tools.[75] His patent, granted in 1819, was not enforced intensively during its initial term. In 1834, however, Blanchard secured from Congress a private act adding fourteen years to his patent. The House Committee on Patents concluded that "the machine is susceptible of being applied to a great variety of valuable uses; and that, by renewing the patent, and thus making it the interest of the inventor to perfect and develop its various capacities, the public may reasonably expect to realize those benefits."[76]

With the extension in hand, Blanchard expanded the reach of his patent enforcement. According to his biographer, Carolyn Cooper, the inventor brought "dozens and dozens" of suits against woodworkers, continuing the campaign after a controversial second congressional extension was allowed in 1848.[77] Reported decisions, most from between 1846 and 1855, trace a line of cases through Massachusetts, Connecticut, New Hampshire, and Pennsylvania. The results suggest that Blanchard benefited almost as much from the liberality of the courts as he had from that of Congress. Surveying the nuances of the technology, Cooper concluded that a generous judicial construction of Blanchard's patent added a creeping breadth to the scope of his rights, allowing them to cover the many related machines that by now crowded the woodworking field.[78]

William Woodworth's planing-machine patent had a shorter duration than the Blanchard patent, lasting "only" from 1828 to 1856, but it evidently cut more deeply. "No patent, it is believed, which has ever been granted in this country, has been so much litigated as this one," remarked Justice John McLean in 1853, on one of the patent's numerous visits to the Supreme Court.[79] Again, the technology in question was a valuable one. The cylinder-head planing machine enabled the rapid production of wooden boards that were flat, smooth, had a uniform thickness, and featured tongues, grooves, and molded features suiting them for floorboards, doors, and other elements of house construction. Labor savings were even greater than for Blanchard's lathe: what had been a journeyman's multiday floorboarding job could now be completed in less than two hours.[80] William Woodworth, a carpenter from Hudson,

New York, conceived a machine along these lines around 1825 and received a patent in 1828. However, lacking the capital to develop or manufacture it, Woodworth quickly parted with his rights: first granting a half share to his local congressman, James Strong; then selling out along with Strong to a syndicate that set out to enforce the patent.

It was this syndicate, led by James G. Wilson, that made the Woodworth patent a phenomenon. Wilson secured two seven-year extensions of the patent, once from the Patent Office and once from Congress. Under the extended term, the syndicate established a network of assignees that functioned as an interregional cartel, setting the price of boards planed on Woodworth-type machines and taking a royalty on each one. By 1852, a hostile congressional committee estimated that $9 million in annual sales of lumber were covered by the scheme and that the owners of the patent had received at least two million.[81] Unsurprisingly, the Woodworth interests were both able and willing to launch hundreds of infringement suits against those who resisted the patent. Wilson claimed to Congress in 1850 that $150,000 had been spent on litigation costs.[82]

§

Alongside the long-lived Parker, Blanchard, and Woodworth grants, a second group of major patent campaigns stamped their mark on midcentury America. These were for more recent inventions: technologies such as the sewing machine, mechanical reaper, vulcanized rubber, and the telegraph.

Samuel F. B. Morse's telegraph invention was closely connected with the patent system from the beginning. Globally speaking, his was not the first practical electric telegraph. Even among American scientists, his contribution to the relevant electrical science would later be debated, minimized, and on occasion scorned. Yet Morse's name quickly became associated with the heroic invention of the new technology: "Professor Morse's lightning."[83] Morse's secret weapon was publicity, wielded by his partners—a series of aggressive politicians and former federal officeholders—and by the Patent Office itself. Henry Ellsworth, the first commissioner of patents under the 1836 Act, had known Morse for years, and seized the opportunity to make Morse's invention and 1840 patent into an advertisement for his new patent administration.[84]

Morse's bid for patent monopoly was part of a complex political and business strategy. The original plan of the inventor and his associates was to sell the patent to the federal government, and most of their early publicity and line building was directed to that end. When congressional purchase failed to materialize, Morse turned to licensing and promoting the construction of telegraph lines.[85] The patent functioned as a tool to manage both competitors and licensees. Litigation inevitably followed and quickly came to turn on Morse's ability to control telegraph technology under a broad reading of his patent. Defendants offered up a variety of telegraph machines purportedly functioning according to different principles.[86] Meanwhile the Morse interests twice obtained reissues of the 1840 patent, in 1846 and 1848, that were designed to make the claims better cover the rival devices.[87] A sequence of far-flung battles in Ohio, Kentucky, and Pennsylvania culminated at the Supreme Court in the case of *O'Reilly v. Morse*.[88] Morse's lawyers pressed the broad claim in Morse's reissued patent for the use of "electro-magnetism, however developed, for marking or printing intelligible characters, signs, or letters at any distances."[89] In a famous decision on the law of patent scope, a divided Court upheld Morse's patent, but invalidated its broadest claim, denying Morse an impenetrable monopoly over the telegraph.

Charles Goodyear's discovery of vulcanized rubber is another classic tale of invention, with just the right blend of genius and serendipity—the rubber mixed with sulfur and accidentally left on a hot stove, producing a newly stable, durable, and useful material. As with Morse, however, the reality behind the legend was of a patent carefully managed to dominance. In the late 1830s, Goodyear was a manufacturer in the struggling New England rubber industry. Through painstaking experimentation, he produced a number of new rubber goods and processes, including the vulcanization method that he began to develop in 1839 and patented in 1844.[90] Goodyear then began granting product-specific licenses to firms making particular rubber goods, shoes, and fabrics.

Goodyear's legal and commercial success depended on two additional efforts. One was an amendment to the patent that broadened its scope. This amendment was made under a provision of the statute that allowed patentees to cancel a defective grant and obtain a "reissue" where inadvertent mistakes existed in the specification or claims.[91] By

1848, commercial rubber could be made by methods that fell outside the terms of Goodyear's patent claim. Goodyear's backers persuaded him to seek a reissue of the patent, ostensibly to correct errors; the reissued version claimed in more abstract terms the application of heat in the curing process, and thus continued to cover the updated product.[92] The other pillar of Goodyear's success was litigation. Relations with both licensees and unlicensed manufacturers were contentious, and Goodyear—who assigned his lawyers shares in the patent—brought more than two hundred suits in the late 1840s and early 1850s.[93] These reached their peak in 1852, with Goodyear's victory in what was universally called the "Great India-Rubber Case" at Trenton.[94] Finally, notwithstanding his victory there, Goodyear was able to persuade the commissioner of patents to extend his patent for an additional seven years, on the grounds that "[n]o inventor before had been so harassed, so trampled upon, so plundered by that sordid and licentious class of infringers known . . . as 'pirates.'"[95]

The final great crucible of patent litigation in the 1840s and 1850s was in new machinery. New technologies of mechanization were led by the mechanical reaper and sewing machine. These complex, multicomponent devices presented a slightly different challenge from the telegraph or vulcanized rubber: a single original inventor would be even harder to nominate, and overlapping or complementary patents would complicate any single claimant's ability to control the field. Even so, a few controversial grants came to the fore.

The reaper was a case where notorious patent campaigns were ultimately unsuccessful. The first practical mechanical reapers were invented in the early 1830s by Obed Hussey of Ohio and Cyrus Hall McCormick of Virginia.[96] Hussey was the first to patent, in 1833, but was less commercially successful. McCormick, on the other hand, became the leading manufacturer of reaping machines in a rapidly growing and potentially vast market. Unfortunately, the industry took off in the late 1840s just as his 1834 patent was reaching the end of its term. McCormick responded by seeking an extension of the patent—an attempt that prompted fierce resistance and became a cause célèbre in Congress around 1850. An early historian reported that "an immense array of political, social, and commercial influence was brought to bear against it by a combination of patent attorneys, rival manufacturers, and agricultural interests; and in the end it was defeated."[97] McCormick opened a series of new fronts with litigation on his subsequent improvement patents in the 1850s. But despite pouring large resources into the suits, they were largely unsuccessful.[98]

The sewing machine, by contrast, featured more enforceable patents than the industry could handle. During the 1850s, it was essentially impossible to manufacture a state-of-the-art sewing machine without running afoul of the overlapping patents covering different features of the device. Thus began the "Sewing Machine War," in which suits and countersuits riddled the industry for years until in 1856 the leading firms combined in a patent pool called the "Sewing Machine Combination."[99] Even here, though, a broad underlying patent emerged. At the heart of the welter of litigation between companies was the 1846 grant to Elias Howe, a penurious independent mechanic who had made early progress toward a working sewing machine, although not a commercially viable product. Howe's success in enforcing his patent across the industry ultimately brought him vast rewards: through licensing and his role in the Sewing Machine Combination, he claimed to have earned more than $400,000 by 1860.[100] In that year, Howe secured a seven-year term extension, which brought his earnings to perhaps $2 million by the time the patent expired.

§

These midcentury attempts to achieve patent monopolies had three consequences for the patent law. The first was a profound effect on patent law and the courts. Mass litigation on a small number of patents dominated the patent business of the federal courts and indeed frequently dominated the caseloads of those courts in general. "The terms of this court are almost wholly occupied in the trial of patent cases," noted one judge in Philadelphia in 1850, while besieged by cases on the Morse telegraph, Woodworth planing machine, and Parker waterwheel patents.[101] The result of swamping the docket in this fashion was that the most-litigated patents accounted for a huge share of the case law. Of the 795 reported patent opinions decided between the 1790 Act and the Civil War, 585 (74%) were accounted for by just 76 patents. The major campaigns were heavily represented: 21 were Parker waterwheel cases, 21 Goodyear rubber, 78 Woodworth planing machine.[102] Many of the leading doctrinal issues of patent jurisprudence—on the law of infringement and claim scope, for example—were worked out in the context of repeat-player, often highly controversial suits.[103]

Substantively, the pressure of the great campaigns began to split the courts' views on the treatment of patents. Patent law had always contained a tension between two perspectives. One saw the grant as a highly desirable reward to meritorious inventors, deserved either because of their great service to society or their natural property in the fruits of their genius, or both.[104] The other regarded patents as a necessary evil, having far more in common with the monopolies of old and warranting strict treatment as a result. Large-scale bids for patent control by self-proclaimed pioneer inventors helped to polarize these positions.

Judges and lawyers upholding the broad grants thus turned to the language of reward, property, and piracy. In the 1839 case of *Blanchard v. Sprague,* Justice Joseph Story delivered what became a classic statement of judicial favor toward patents, contrasting an old English suspicion of the grant ("adopted upon the notion, that patent rights were in the nature of monopolies, and, therefore, were to be narrowly watched") with America's "far more liberal and expanded view of the subject . . . a natural, if not a necessary result, from the very language and intent of the power given to congress by the constitution."[105] Therefore, Story reasoned, patents were "clearly entitled to a liberal construction."[106] Judges upholding the Woodworth, Morse, and other powerful patents repeatedly returned to this theme, adding frequent notes of recognition for property rights. "The invention which is set forth in letters patent belongs to the inventor as rightfully as the house he has built or the coat he wears," wrote the judge in one Woodworth case, adding that "[i]t would be a reproach to the judicial system if an ownership of this sort could be violated profitably or with impunity."[107]

Conversely, judges more skeptical of patent rights—especially those steeped in the Jacksonian suspicion of monopoly—addressed the patent right as a government-granted "franchise," of the type that the courts were then widely working to construe narrowly against their holders.[108]

Rather than seeking to give patents a liberal construction, these judges believed that the patentee "ought to state distinctly what it is for which he claims a patent, and describe the limits of the monopoly."[109] In the 1850s, this divide split the Supreme Court in a series of cases now regarded as formative in the modern patent law. In *Winans v. Denmead* (1853), for example, a 5–4 majority of the court endorsed the "doctrine of equivalents," allowing infringement beyond the scope of the literal claims, whereas a minority insisted on strict adherence to the letter of the patent. In that same term, a slight shift in votes led to the majority in *O'Reilly v. Morse*, striking down a broad claim as a patent for a mere principle—a decision that would become a foundational restriction on claim scope.[110] What has come down to us as a "golden age" of patent lawmaking by the courts was really a scene of chronic conflict.[111]

Conflict was central to the second consequence of the great patent campaigns: the close entanglement of patent law with the political branches. The middle nineteenth century has historically been portrayed as a judicially driven period in patent law, with Congress remaining on the sidelines.[112] Although this depiction may be true regarding general patent legislation, it misses the large amount of private action in patent matters sought and granted through Congress. Congressional interventions, above all in the form of private bills extending patents, were clearly highly influential and highly controversial in the politics of patents.

On the one hand, there was lobbying. Vast sums were poured into extension battles. A congressional inquiry of 1854, appointed to investigate charges of bribery surrounding the attempted extension of Samuel Colt's revolver patent, painted a lurid picture of the "agents, attorneys, and letter-writers" employed to bombard legislators in patent extension cases.[113] "Costly and extravagant entertainments" were laid on for "ladies and Members of Congress and others" in support of extension bills.[114] The "most effective agents" available for hire were the credentialed correspondents of the daily press, whose access to the House floor was supposedly contingent on a pledge not to lobby, but who in practice were employed en masse by the backers of "railroad, patent, and other schemes."[115] The Colt investigation turned up no hard evidence of bribes to congressmen, but its hearings and report managed to convey the sense that these were universal.

On the other hand, opposition to patent extensions produced genuine popular mobilizations. "Remonstrance after remonstrance" was sent to Congress against a further extension of the Parker patent in 1854, coming "from Maryland, and Pennsylvania, and New York, and Maine, and Indiana, and indeed from almost every state in which mills are used," and each "signed by hundreds of individuals."[116] In Philadelphia, a "mass meeting" of lumbermen and carpenters was held in 1850 to arrange resistance to the proposed Woodworth patent extension; the Pennsylvania legislature was one of a number that passed resolutions against congressional action.[117] Similar agitation greeted the lobbying of McCormick, Goodyear, and Howe. The bitterness of these battles easily rivaled the publicity generated by litigation. In 1861, anger with selectively protracted patent rights led to the end of administrative extension practice: the commissioner's power to extend grants by seven years was ended, and the standard patent term lengthened to seventeen years instead.[118] Attempts to obtain congressional extensions dwindled after most of the major efforts in the 1850s were turned back. Supporters and opponents of patents would continue to turn to Congress and the executive, however, albeit in different ways.

The final consequence of the midcentury patent wars was the involvement of the elite bar. The litigation campaigns of the 1840s and 1850s generated expensive, far-flung, and long-running litigation, which drew legal talent from the top echelons of law and politics. The Great India-Rubber Case pitted the famous advocate Rufus Choate against the venerable Daniel Webster, then secretary of state, who provided his legendary talents for the final time in return for an enormous fee.[119] Of similar stature, sometime U.S. senator and attorney general Reverdy Johnson represented the Woodworth patent owners and Cyrus McCormick.[120] New York's governor, William H. Seward, also represented the Woodworth interests. Ohio senator Salmon P. Chase argued for Morse. The money to be made from these cases, along with their location in the rarefied federal courts—and the advantages of having a prestigious advocate prevail on the court in case the judges struggled with technical detail—ensured that patent litigation swept in an appreciable proportion of the antebellum legal-political elite. Three members of Abraham Lincoln's wartime cabinet—Secretary of State Seward, Secretary of the Treasury

Chase, and Secretary of War Edwin Stanton—had appeared in high-profile patent business, and Lincoln himself had played a walk-on part (and reportedly earned the second-highest legal fee of his career) in the McCormick reaper litigation.[121]

Even as antebellum patent litigation drew in the existing legal establishment, a new and unusual development occurred: a class of more specialist practices began to emerge. For most of the nineteenth century, there were some near-universal truths of legal practice: law was intensely local, built on community networks and familiarity with local courts. Lawyers were predominantly generalist, taking cases in a range of fields more determined by the composition of their client base than by their substantive areas of knowledge. Specialization only began to emerge around 1870, led by those parts of the profession closest to the insurance industry and to industrial and financial corporations. Patent law was, in many ways, an exception to these patterns, emerging as a distinctive field of endeavor before the Civil War. The reasons were not solely related to technical knowledge, although that played a role. Instead, they had much to do with the institutions at the heart of the patent law: the Patent Office and the federal courts.

After the 1836 Patent Act, a small group of specialists sprang up around the Patent Office. Charles M. Keller, the first examiner appointed under the act, acquired legal training and began practice in New York in the 1840s; Charles Mason, who served as commissioner of patents from 1853 to 1857 later cofounded the prominent patent law firm of Mason, Fenwick & Lawrence.[122] Clustered in Washington, D.C., and the major metropolitan centers, a growing group of patent attorneys focused their practice on drafting and preparing patent applications.

The great litigation campaigns created their own class of patent litigator. William H. Seward was a key transitional figure. Shortly after Seward's first term as governor (1839–1843), James G. Wilson retained him in the massive and bitter campaign to impose the Woodworth rights on the woodworking trade. The task transformed Seward's law practice from its upstate New York origins to a regional and even national endeavor. Demand for Seward's patent expertise took him to the federal courts of New York City, Philadelphia, Baltimore, Washington, and as far west as St Louis.¹²³ His practice, a variety of overlapping partnerships that eventually resolved into the firm of Blatchford, Seward & Griswold, took on as clients McCormick, Howe, Morse, the great steam engineer George H. Corliss, the leading railroad inventor Ross Winans, and Charles Goodyear's principal opponent, Horace Day.

Gradually, a cadre of elite national patent litigators took shape: figures such as George Gifford, the leading patent lawyer in New York from the 1840s onward, who appeared for Morse in the telegraph cases and was a leading figure in the litigation around the sewing machine; and George Harding, his counterpart in Philadelphia. Both Gifford and Harding were occasionally dubbed the "father" of the specialty.¹²⁴ After the Civil War, they would be joined by a new generation of patent specialists— increasingly, scientifically trained lawyers who were able to devote their practices to the massive wave of patent business in the late nineteenth century. It was to these men that later inventors would owe their fates.

§

Fittingly, a good illustration of the extent of patent litigation in mid-nineteenth-century America hangs in the National Portrait Gallery, the current occupant of the grand old Patent Office building. The group portrait *Men of Progress* was painted in 1862 by Christian Schussele, a prominent Philadelphia artist of French-Alsatian origin. It depicts nineteen men, described as "the most distinguished inventors of this country, whose improvements . . . have changed the aspect of modern society, and caused the present age to be designated as an age of progress."¹²⁵ One striking aspect of this pantheon of recent greats is the intensity of patent litigation that lay behind it. At least eleven of the nineteen were involved as plaintiffs in significant patent cases on one or more of their signature inventions.¹²⁶ Five of them were among the most prolific and consequential patent litigants of their time: Thomas Blanchard, Charles Goodyear, Samuel Morse, Cyrus McCormick, and Elias Howe. Their presence among the "Men of Progress" stands for a key feature of nineteenth-century patent law: the close relationship between valuable technology, patent litigation, and publicity.

The last third of the nineteenth century saw patents and patented inventions emerge as a still greater force in America's industrial development.

The patent system both drew on and contributed to the idea of the Great Inventor. Of the pantheon of American inventors assembled in the portrait *Men of Progress*, more than half had engaged in significant patent litigation over their inventions. Christian Schussele, *Men of Progress* (1862), National Portrait Gallery, Smithsonian Institution; transfer from the National Gallery of Art; gift of the A. W. Mellon Educational and Charitable Trust, 1942. NPG.65.60.

Over 600,000 patents were issued in the United States between 1865 and 1900, more than ten times the number created in the seventy-five years prior to the end of the Civil War, and more than twice as many as were granted by any other country.[127] The exploitation of these rights became a signature theme of the period. Sprawling legal campaigns accompanied many great inventions of the age, from electric lamps and telephones to less esoteric articles such as barbed wire and baking soda. The battles over these technologies would equal the midcentury patent wars in political intensity and surpass them in number and scale.

They would also pose new questions. Industrial developments in the later nineteenth century forced back into focus the relationship between intellectual property and monopoly. Monopoly, in this new context, did not refer to an illegitimate privilege but to an economic phenomenon: the growing concentration of market power in the hands of large corporations.[128] The legal apparatus of the patent system had scant formal means to respond to these changes. The remit of the law remained the same: to identify the first and true inventor and to match the scope of protection to the original inventive contribution made and claimed. Eventually, this conception of the patent as a reward to individual genius would seem at best complicated and at worst dysfunctional in the face of its use as an instrument of market power.

At root, however, the first question of the patent remained the same as it had ever been. What had the claimant invented, and when had he invented it?

CHAPTER 2

Acts of Invention

THE HISTORY OF THE TELEPHONE did not begin with an invention. Instead, it began when a small group of entrepreneurs, including Alexander Graham Bell, formed a company for the purpose of seeking improved methods of telegraph transmission. These men plunged into a race to develop sound-based electrical signals, a contest which led unexpectedly to the transmission of human speech. What happened next—the commercialization of the telephone as a disruptive stand-alone technology, independent of the telegraph—was not inevitable. Nor was it the result of visionary choices by an individual inventor. Instead, the fate of the new device depended heavily on the character of the Bell enterprise: a high-tech start-up whose strategies were driven by investor relations and the exploitation of patents.

From the beginning, the inventive effort that produced voice transmission was conducted with the telegraph in mind. Telegraph technology was by far the most significant existing application of electricity in the mid-1870s, dwarfing the early research then going on into electric lighting and power. The telegraph business was dominated by Western Union, which had consolidated the major lines in the aftermath of the Civil War and in the process had become one of the largest corporations in the United States. Smaller specialized firms served the markets for private lines, printing telegraphs, stock reporting, and alarm services.[1]

Within the ecology of telegraph companies and the machine shops that supplied them, patent rights held a prominent place. The centrality of patents was not entirely new, of course: Samuel Morse's patent had helped to shape the telegraph business in its early stages. But the 1870s saw industry leaders embracing innovation as a business strategy and using patents systematically to manage and control the process. In an age before corporate research and development, this approach alone was a significant organizational step. One notable pioneer was the Gold and Stock Telegraph Company, a provider of stock ticker service in New York, which retained selected inventors and sought out promising new patents in the innovative field of intracity communications. Western Union itself became a technologically progressive outfit under William Orton, the company's president from 1867 to 1878. Orton maintained close contact with inventors, including Thomas Edison, a young electrician associated with the Gold and Stock Company.[2]

Arguably the richest seam of telegraphic progress being mined at the time was in multiple-transmission technology, and it was this enterprise that would break through into voice communication. In 1868 the Boston-based inventor Joseph Stearns successfully demonstrated his duplex telegraph, which enabled a telegraph wire to carry two signals simultaneously in opposite directions. Duplex working approximately doubled the capacity of an existing system with only a small increase in plant, leading Western Union's Orton to consider Stearns's device the most important contribution to the industry since Morse.[3] Thomas Edison then developed a quadruplex system that multiplied transmission capacity yet again. Although Edison had conducted his research under an arrangement with Western Union, the telegraph company had failed to secure his patent rights. Edison promptly sold his quadruplex patent to the Atlantic and Pacific Telegraph Company, an insurgent competitor controlled by the tycoon Jay Gould. For this act, Orton and his allies dubbed Edison a "Professor of Duplicity and Quadruplicity."[4] Western Union and the Atlantic and Pacific engaged in bitter litigation over the quadruplex patent, which ended only when Western Union paid a ransom to buy out the Gould firm in 1877.[5]

Research on the electrical transmission of sound during the mid-1870s was the offshoot of this high-stakes, patent-oriented search for improvements in telegraphy. The hoped-for "harmonic telegraph" represented a further advance on the quadruplex system. It promised to cram even more traffic into the wires by using signals generated simultaneously by different acoustic pitches, which could then be unscrambled into separate messages by the receiver. Early progress in this direction was made by the electrical inventor and telegraph equipment manufacturer Elisha Gray. Later, Western Union drew Edison into the chase under an exclusive contract.[6] The scale of the rewards available—Western Union had paid Stearns $50,000 for the duplex, and Edison had sold his quadruplex rights for $30,000—also gave plenty of encouragement for research outside the sponsorship of established firms.[7]

Alexander Graham Bell came to acoustic telegraphy from speech therapy, the profession he shared with his father. In 1870, at twenty-three, he moved with his parents from London to Canada, and the following year relocated to Boston as a teacher of the deaf. Amid Boston's flourishing scientific culture, Bell revived his long-standing interest in electricity and especially in the application of electricity to acoustic research. Press coverage of Joseph Stearns's duplex telegraph inspired him to think about how sound might be used in multiplex telegraph signaling. Bell began to experiment with harmonic telegraphs in 1872, building his own electromagnets and running wires from his rooms to those of a sympathetic neighbor.[8]

As Bell himself pointed out, his occupational experience began with sound and led him to electricity, whereas his eventual competitors had moved in the opposite direction. This made him, in some sense, an outsider. Although there may have been differences in the approach he took toward technical problems, Bell was by no means excluded from the mainstream of inventive production.[9] Invention and patenting were not primarily an internal function of firms, and technological leaps were not yet the preserve of the big battalions in research and development. Even professional inventors such as Elisha Gray and Thomas Edison were essentially independent operators. Most highly productive inventors, in fact, assigned their patents to several different firms over the course of their careers. Only at the end of the nineteenth century did large companies—Bell Telephone included—turn inward to industrial research rather than expending their energies on the outward search for others' inventions.[10]

Electricity was, to be sure, a relatively "scientific" field, increasingly informed by the work of trained physicists.[11] But the state of the electrical arts in the 1870s did not preclude major advances by independent inventors. Bell's amateur status could readily be offset by the open scientific and technical networks that Boston afforded. Scientific engineering in the city centered on the Massachusetts Institute of Technology, where Bell attended lectures on acoustical physics and electricity and had several academic consultations. For practical engineering skills and materials, he could turn to the world of machine shops that incubated much telegraph innovation.[12] One institution in particular helped to lower the overhead costs of electrical research in Boston: Charles Williams's workshop on Court Street. Williams had supplied equipment and laboratory space for the inventors Moses Farmer, Joseph Stearns, and Thomas Edison. He provided similar facilities to Bell, along with the assistance of his young mechanic, Thomas Watson.[13]

Even in this auspicious climate for telegraph innovation, Bell by himself might have struggled to make anything of his experimental work. Crucially, though, his inventive prospects attracted investors. Bell's speech work brought him into contact with the lawyer Gardiner Greene Hubbard, whose deaf daughter Mabel was one of Bell's pupils (and would later become his wife). Hubbard had worked as a patent attorney early in his career, organized a number of utility enterprises, and had become deeply involved in telegraph matters. Since the 1860s he had been a high-profile critic of the Western Union monopoly and had lobbied Congress to charter a Postal Telegraph Company that would compete with and undercut the allegedly oppressive incumbent.[14] Seeing in harmonic multiple telegraphy a new technological weapon against Western Union, Hubbard became Bell's chief patron and promoter. Thomas Sanders, a leather merchant from Salem, Massachusetts, and father of another deaf student, joined the two in forming the Bell Patent Association, which proposed to divide equally any telegraph patents forthcoming from Bell's research.[15]

Hubbard's and Sanders's investment enabled Bell to continue his telegraph experiments and bring them to patentable form. At the same time,

Hubbard's sponsorship gave Bell access to heavyweight legal expertise, in the form of the Washington lawyers Anthony Pollok and Marcellus Bailey. These two men were among the capital's leading patent attorneys. Bailey was the son of a prominent abolitionist; he had served as an officer with the U.S. Colored Troops during the war and continued to be known as "Major" Bailey for years afterwards.[16] His partner, the Hungarian-born, French-educated Pollok, cut an urbane, cosmopolitan, and conspicuously wealthy figure. He impressed Bell deeply, not least by hosting the inventor at his "palatial residence . . . the finest and best appointed of any in Washington" and by leading Bell on a circuit of dignitaries that included "Mrs. Bancroft (the wife of the historian) . . . Prof. Henry of the Smithsonian . . . Sir Edward Thornton and the members of the other Foreign Embassies."[17] Pollok was Bell's lawyer, but he was Hubbard's confederate. As an associate of Hubbard's in the Postal Telegraph effort, he shared the goal of placing Western Union on the defensive using the new acoustic technology.[18]

Pollok and Bailey were among the elite of the new patent law establishment, men who were busy growing rich on the boom in patenting that followed the Civil War. Their skills in patent drafting and management were undoubtedly formidable. More than that, though, Pollok and Bailey were men of influence, whose networks of intelligence extended deep within the Patent Office. According to a fellow lawyer, Pollok had once been heard "boasting that he could control the Office."[19] Bailey's assets, fatefully for Bell, included a financial relationship with the Patent Office's principal examiner of telegraph inventions.[20]

By the early spring of 1875, Bell had developed several elements of a system of multiplex telegraph transmission. At this point he knew he was not alone in the field. Hubbard and others regularly reported on the activities of the Chicago inventor Elisha Gray, who began to file a series of patent applications relating to harmonic telegraph inventions during 1874. Bell came to believe it "a neck and neck race between Mr. Gray and myself who shall complete our apparatus first."[21] Gray's having filed earlier patents was not an insuperable obstacle. As Hubbard pointed out, priority of invention rather than priority of filing was the key to a valid patent: "A Patent is of no value provided another made the invention at an earlier date and the issue of a Patent to A does not prevent B from subsequently obtaining a Patent for the same invention if he was the prior inventor."[22] In addition, Pollok and Bailey came up with a strategy for frustrating Gray's patent efforts. Their method was to exploit the Patent Office procedure known as an "interference"—a formal hearing conducted to establish who, between competing applicants, was the true first inventor. As Bell prepared to file for his first harmonic telegraph patent in February 1875, Pollok and Bailey had him divide his submission into three separate applications. They correctly predicted (perhaps thanks to intelligence from inside the Patent Office) that one or two of Bell's three filings would tie up Gray's pending submissions in interference proceedings, while the third application would slip through unscathed.[23]

The legal maneuvering over harmonic telegraphy was matched by jockeying for commercial position. While their respective applications were pending, both Bell and Gray demonstrated their devices to the buyer of first choice, Western Union's President Orton. Gray appeared to come out ahead of Bell in these exchanges, partly by offering a rather more finished product and partly because Orton was reluctant to benefit his political adversary Hubbard.[24] Having opened negotiations with both rivals, Orton then put Thomas Edison on retainer to produce his own version.[25]

That summer, Bell, to the frustration of his partners, allowed himself to be distracted from the guaranteed rewards of multiplex telegraphy by a more radical possibility: the transmission of speech itself. Bell's key theoretical recognition, which had been forming in his mind for almost a year, was that complex sounds could be transmitted by the use of a continuous and fluctuating ("undulatory") current rather than by the intermittent electrical current used by the telegraph. Experiments with Thomas Watson in the Williams shop in June and July 1875 revealed the sensitivity with which sound could be reproduced—first that of a plucked reed, then inarticulate vocal noises. Bell was giddy at the possibilities of this "[g]rand telegraphic discovery," telling a friend that he hoped with modifications to "distinguish—in the echo—the 'timbre' of the sound. Should this be so—conversation *viva voce* by telegraph will be a fait accompli."[26] Conversation had to wait, though, as teaching obligations and pressure from Hubbard to work on the harmonic telegraph prevented any further experiments with voice transmission.

At the beginning of 1876, Bell finally turned his attention to a patent for the undulatory current idea. During January he prepared two versions. One was for a British patent application, which he planned to have filed in London by a family friend, the Canadian newspaper publisher George Brown. The other was drawn up for the U.S. Patent Office. Bell apparently drafted the specifications himself, without input from his lawyers, before having Pollok review them and pronounce himself "very much pleased."[27] Hubbard and Pollok then pressed for expeditious filing because there was "much excitement in this branch of Telegraphy" and they were worried that Bell would allow word of his method to leak.[28] However, Bell's arrangement with George Brown required a delay. Brown and his brother had agreed to pay for a share in any British patents, but had insisted that these be filed before the American applications to avoid any problems of prior publication under British law.[29] There things stood at the end of January 1876. Bell notarized the American application in Boston on January 20 and sent it to Hubbard in Washington. Brown sailed for England on the 26th, promising to cable word when the British patent was filed.

On February 14, having held the specification for some weeks, Hubbard suddenly acted. Bell's application was hand-delivered to the Patent Office in the morning. At some point that same day, Elisha Gray submitted his own description of "the art of transmitting vocal sounds or conversations telegraphically, through an electric-circuit."[30] In popular histories, this is a crucial moment: Bell beating Gray to the Patent Office by mere hours, and going on to claim the prize of telephone inventorship. There were some implications to the timing of the submissions, as we shall see. Yet as a general legal matter, the order of filing did nothing to determine the winner of a patent race: a valid patent under U.S. law went to the first actual inventor, rather than the first to register a patent application. Whenever two pending patent applications claimed the same subject matter, the Patent Office would suspend both and resolve inventive priority in an interference hearing, regardless of when the filings came in through the door.

Bell's and Gray's near-simultaneous filings are also often cited as history's most famous example of simultaneous invention, but the chronology of their inventions was nowhere near so neat. Bell's patent was based on his slowly evolving theory of the undulatory current and on his vocal-sound experiments with Thomas Watson more than seven months earlier. Gray had periodically considered the voice-transmission possibilities of his harmonic telegraphs, and had at one point elicited vowel sounds from one of his devices, but had resolved not to spend time and money pursuing a prospect without immediate commercial application.[31] The document filed by Gray reflected as much: it was not a full patent application but a "caveat," a statement of conception and intention to develop an invention, used to create evidence of priority. Neither man had yet transmitted intelligible speech.

Stage-worthy drama aside, what the events of February 14 really reveal is the deep game that Bell's lawyers were playing at the Patent Office. Many of the relevant details were subsequently uncovered during the patent litigation of the 1880s; others by twenty-first-century sleuths who have meticulously reconstructed the sequence of events.[32] From these sources, an unflattering picture of behind-the-scenes activity emerges. The first suspicious act was the sudden submission of Bell's patent, which inexplicably broke Bell's agreement with George Brown and took place without Bell himself being warned or consulted. This alone suggests that Hubbard or the lawyers had gained advance knowledge of Gray's plan to file. There then followed a rapid succession of moves within the Patent Office, many of which were irregular. Officials initially declared an interference hearing, standard in such cases, to determine priority of invention between Bell and Gray. However, they then summarily withdrew the order, after Bell's lawyers argued that their client's application had arrived before Gray's caveat. Unlike a full patent application, a caveat would trigger an interference hearing only if filed before the full application with which it conflicted. Happily for Bell, the Patent Office had recorded the exact time of delivery of his application, although it was not the usual practice to do so. Even more fortunately, the acting commissioner of patents, in withdrawing the interference, accepted Pollok's arguments that the time of day should matter. In a similar case reported only three weeks earlier, he had ruled that the hour of submission had no bearing and that the date alone mattered.[33]

Because of these decisions, Bell's patent application escaped the months-long or even years-long delay and uncertainty that an interference battle could bring. The irregularities did not end there. Bell was permitted to make some amendments to his pending application—a permissible move on its face, but one that suggests the Bell interests had illicit knowledge about yet another potentially interfering Gray application. Finally, the Patent Office issued Bell's patent, in this case and in all his other applications, exceptionally rapidly by its usual standards.[34] Some of the important background to these events subsequently became part of the public record, notably that Zenas Fisk Wilber—the patent examiner who handled Gray's and Bell's applications—was heavily in debt to Bell's lawyer Marcellus Bailey.[35] Many other details of who knew what, when, and how remain obscure.

On March 7, 1876, Bell received the fruits of his and his lawyers' labors. U.S. patent number 174,465, for "Improvements in telegraphy," would become the fundamental patent that controlled the telephone. On its face, the patent related primarily to harmonic telegraphs. "My present invention," the specification explained, "consists in the employment of a vibratory or undulatory current of electricity in contradistinction to a merely intermittent or pulsatory current, and of a method of, and apparatus for, producing electrical undulations upon the line wire."[36] The patent went on to detail the benefits provided by the undulatory current, all of which involved increasing the capacity or speed of a telegraph circuit, and none of which mentioned speech transmission. Only two references to voice action—and even then, not articulate speech—appeared in the document. One passage noted that "the armature *c*, Fig. 5, may be set in vibration . . . by wind. Another mode is shown in Fig. 7, whereby motion can be imparted to the armature by the human voice or by means of a musical instrument." This language accompanied a drawing that depicted, in simple schematic form, the instrument on which Bell and Watson had reproduced vocal pitch in July 1875. The other reference lay in the fifth claim, or formal statement of invention, which asserted rights over "[t]he method of and apparatus for transmitting vocal or other sounds telegraphically, as herein described, by causing electrical undulations similar in form to the vibrations of the air accompanying the said vocal or other sounds, substantially as set forth."[37]

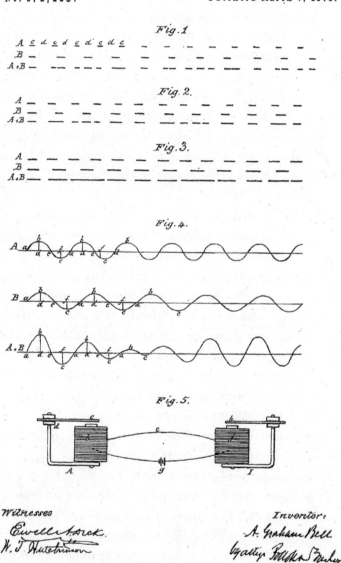

Drawings from U.S. Patent 174,465, issued to Alexander Graham Bell March 7, 1876, for "Improvements in telegraphy." The diagram marked "Fig. 4" depicted Bell's key conception of the undulatory current. "Fig. 7" showed, in schematic form only, an instrument capable of transmitting vocal sounds. Courtesy of the United States Patent and Trademark Office.

Acts of Invention ～ 45

No. 174,465.

A. G. BELL.
TELEGRAPHY.

2 Sheets—Sheet 2.

Patented March 7, 1876.

Fig 6.

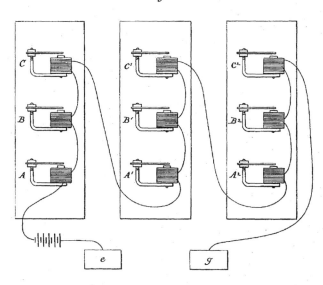

Fig. 7.

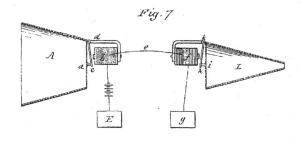

Witnesses
Ewell Hoick
N. J. Hutchinson

Inventor:
A. Graham Bell
by atty Pollok & Bailey

One could not know, on March 7, 1876, that this claim would govern the transmission of speech by electricity. And as the future controlling patent of the telephone, number 174,465 also had one other notable feature: its inventor had never transmitted an intelligible word. Proof of prior working was not a requirement of the patent statute; technically, all a patentee had to do was provide a "written description" of the invention sufficient "to enable any person skilled in the art or science to which it appertains . . . to make, construct, compound, and use the same."[38] Fortunately for Bell, the device sketchily described in his patent could—it later transpired—be made to work as a telephone. Not that achieving such a result was an easy matter. As late as 1879, Bell would find himself, on the eve of the first major patent litigation, huddling with his main expert witness to "try certain experiments to see whether the form of telephone shown in my first patent will work."[39] In any event, Bell's actual breakthrough used different methods. On March 10, 1876, three days after receiving his crucial patent, Bell spoke and Watson heard the first sentence sent by telephone. The device they employed looked little like the apparatus in patent 174,465, although it did bear a resemblance to the machine described in Gray's caveat.[40]

Successful speech transmission put Bell's telephone work on a different footing. Although his partners in the Bell Patent Association continued to hope for a multiplex harmonic telegraph, they warmed to the telephone in the months after March 1876. Bell began to supplement his experiments with public lectures and touring demonstrations of the telephone. In July, he achieved his signature triumph at the Philadelphia Centennial Exhibition: a successful test that attracted endorsements from no lesser figures than the emperor of Brazil and the eminent British scientist Sir William Thomson. The remainder of the year was spent testing, refining, and improving the device.[41]

In January 1877 Bell secured his second telephone patent, number 186,787. In this patent, the inventor explicitly laid out a system of "electric telephony" and described his working telephone, now vastly more practical than the flimsy device sketched in the first grant.[42] Bell's design was what he called a "magneto" telephone. The receiver contained a permanent electromagnet that vibrated a metallic "armature," or diaphragm, to produce a sound corresponding to that made at the transmitter. The magneto would remain the basic principle of the telephone receiver for decades.

Bell's second patent occupies a curious place in this story. Because it covered the basic form of receiver, it has been regarded ever since as one of Bell's two fundamental patents for the telephone. In litigation, the two were a pair: every time the Bell Company sued under the original patent of 1876, it also alleged infringement of the 1877 grant. Yet patent number 186,787 played a very distant second fiddle. The great judicial decisions on the scope of Bell's first patent barely mentioned his second, except to note it valid and infringed. Bell's lawyers never made grand claims on its behalf, describing its content as a "mere mechanical improvement" on the instrument in Bell's 1876 patent.[43] By virtue of its later issue date, patent number 186,787 did have a brief moment in the sun. During the eleven months after Bell's first grant expired in 1893, the Bell Company brought and won a handful of infringement suits under the second patent alone. But the judges in those cases simply referred to its long history of being upheld by the courts.[44] Until this second, supplementary patent expired on February 1, 1894, no telephone in the United States would be free of Bell's legal claims. Yet it is impossible to say whether, in the absence of the all-conquering 1876 patent, Bell's magneto receiver would have held any broad sway over telephone technology—the claims were simply never tested in that way.

§

Therein lies the rub. For all the justifiable historical attention paid to the process of issuing Bell's patents, the grant of his rights was not the end of the story. Patents were and are highly malleable legal artifacts, capable of being constantly shaped and reshaped. The first opportunity to demarcate the scope and content of a patent comes during the drafting of an application, when a patentee chooses how to describe the invention and which features should be formally claimed as original. But the owner of the grant also has opportunities to redefine the content of the patent at a later date. One route for doing so is through administrative reissue, a tactic that was heavily used (and abused) in the middle nineteenth century. Another is through the interpretation of the patent put forward in litigation. Alexander Graham Bell may have received his patents in 1876 and 1877, but the question of what he had "invented" as a legal matter was yet to be determined.

Also yet to be decided was the question of whether Bell's patent would make any significant difference to the development of the telephone. Even where the scope of the grant remains perfectly stable, a patent's practical implications depend to a very large extent on its owner. Because patentees can choose how to exploit their rights, even a patent comprehensive enough to cover a particular technology completely (a "legal monopoly") need not create concentrated supply (an "economic monopoly"). Some inventors position themselves as manufacturers. Others license their inventions. Charles Goodyear, for example, marketed the rights to his hardened rubber and allowed others to produce the boots, clothing, and industrial products that used it.[45] A patentee's incentives to maintain exclusivity or to license freely depend in part on his position in the market: whether, for example, he is a "user-inventor" who might closely guard the invention for his own competitive advantage, or a "manufacturer-inventor" with an incentive to diffuse the device widely so as to earn maximum sales or licensing revenue.[46] The practicalities of the market itself also play a role. In the nineteenth century it was difficult, if not impossible, for any single firm to satisfy regional (let alone national) markets for its product. Patentees thus commonly licensed others to make their products, or assigned territorial rights for their inventions, even if they hoped to do some of the manufacturing themselves.[47]

Similar factors help to explain why patents matter more to competition in some industries than in others. The ability of a patentee to extract economic rent from his property depends on a host of factors related to his position in the market, including power relations between different levels of the supply chain and relations between the patentee and the leading firms in the industry. America's late nineteenth-century railroad companies, for example, had little interest in competing through technological differentiation. As a result, they worked to minimize the influence of patents by pooling their own grants and colluding against external patentees.[48] This approach contrasted sharply with the jealous and aggressive patent strategies of their contemporaries in the new electrical industries, firms that were formed to exploit new patented products or to generate potentially controlling grants of their own.[49] The divergence between "soft" patent regimes dominated by cross-licensing and "hard" ones defined by exclusive use and litigation had little to do with the inventions themselves. Instead, it depended on industry structure and the strategies of exploitation that patentees were in a position to pursue.

The formation of a proprietary interest in the telephone began when Bell and his investors came together to seek patentable inventions in telegraphy. It culminated in a telephone industry based on proprietary networks protected by legal barriers. Patent monopoly over the telephone owed its existence and form to the path that the technology followed to market. That path had two key waypoints. First, voice communication emerged as an independent industry rather than an auxiliary to the telegraph. Second, the patentees—lacking any major assets other than their legal rights—adopted a business structure constructed around maximal exploitation of the patents. In particular, they hitched their ambitions to the promotion of telephone service rather than construction of the patented instruments alone. Not only the telephone instrument but the whole edifice of telephony, from local exchanges to thousand-mile lines, would fall under the patents as a result.

The path from receipt of the first patent to the formation of a company for exploiting it was not straightforward. Late in 1876 or early in the following year, Hubbard offered Western Union the rights to the telephone for $100,000, only to have Orton turn him down.[50] A series of other negotiations over the Bell patent also fell through. These included a proposal to form a company with the governor of Rhode Island and Elisha Converse of the Boston Rubber Shoe Company, both sometime business associates of Hubbard. The manufacturer Peter Cooper demurred after deciding that the likely patent litigation would significantly decrease the value of any investment.[51] In July 1877 the partners struck out on their own, forming an unincorporated company and dividing five thousand shares among Hubbard, Sanders, Bell, and Watson. All Bell's telegraph and telephone patents, present and future, were assigned to Gardiner Hubbard as trustee.[52]

The Bell Telephone Company of 1877 may have laid claim to legal ownership of the telephone, but it did not attempt to conduct the business itself. Manufacturing at least lay close at hand because the company bought its telephones exclusively from Charles Williams's workshop next door on Court Street.[53] But the company turned to others to promote and construct telephone lines and to provide service. The local entrepreneurs who did so, and who pocketed the price from their customers, were linked to the Bell Company only by exclusive area franchises and contracts for the supply of telephone instruments. In no meaningful sense were they under the control of Bell Telephone. Hubbard and his colleagues had not much more than a hortatory influence on the dispersed agents, especially as the telephone spread west and south.

Even so, it would have been clear to every customer that the telephone was a proprietary device. The patentees worked to ensure that every telephone in the United States was a Bell telephone by retaining ownership of the instruments and merely leasing them to subscribers. The revenue from instrument rentals, of which agents kept a commission of some 20 to 50 percent, provided the parent company's income.[54] According to the patent lawyer James J. Storrow, whose long association with the company began shortly afterward, the telephone leasing plan was inspired by the success of the McKay Shoe Machinery Company.[55] Both Gardiner Hubbard and Bell's general counsel Chauncey Smith had represented the McKay Company, which dominated the supply of shoe machinery in America from its Massachusetts base. Having paid $70,000 for a shoe-sewing machine patent in 1859, McKay leased machinery to shoemakers for a small installation charge and a royalty of up to three cents for each pair of shoes manufactured. By 1876 annual revenues from this arrangement were over half a million dollars.[56] This model appealed to the Bell partnership for financial, technical, and legal reasons.[57] It provided a steady flow of income to the capital-starved telephone company by spreading revenues and softened up the market by lowering entry costs for cautious customers of the still-unproven technology. For customers, the rental mechanism also provided a solution to the problem of guaranteeing quality because it gave the company an interest in maintaining and continuously improving the instruments.[58]

The vertical structures in the early telephone industry—local franchising and instrument leasing—decided the role that the patents and the patent-owning company would play. The practice of leasing rather than selling instruments cast the Bell Company's product as an ongoing service instead of a durable good, and in doing so gave the patent-holding company an interest in the conditions of network operation downstream. At the same time, exclusive area franchises gave licensees their own local stake in a Bell patent monopoly and effectively recruited them to guard against infringements. As a result of these ties, patent rights became the fundamental shaping influence on the competitive structure of telephone service.

The telephone industry could have developed in other ways. It is easy to imagine an alternative situation in which the companies that built telephone networks were either independent of the patent owners or strong enough to defy them. In this scenario, existing telecommunications operators—Western Union being the obvious candidate—might have dominated the provision of telephone systems, purchasing the instruments they deployed and thus viewing returns to patentees as an unwelcome cost. This kind of regime emerged in the American railroad industry, where the leading companies declined to engage in patent contests with one another and instead cooperated against the patent-wielding manufacturers of railroad equipment. A similar situation governed the proprietary fate of the telephone in a number of countries. Germany's Imperial Postal Telegraph authority, for example, immediately assimilated the telephone as a feeder system to its existing network and then used its procurement policy to foster competition among rival instrument suppliers.[59] In the United States, however, the patentees engaged in the supply of telephone service, rather than simply the sale of equipment, and consequently had a competitive interest in distribution.

Competition was not long in coming. Having failed to set the industry off on an alternative path by commercializing the Bell patent as part of its telegraph empire, Western Union entered the fray with a rival set of patents. Whether or not the telegraph company had ever seriously weighed the $100,000 offer from Hubbard, Western Union's experiences with other new message services in the 1870s gave a reassuring record of patents bought out or outflanked-to-order by the prodigious Edison.[60] Doubtless confident of his ability to drive the Bell concerns into capitulation or compromise, Orton supervised the assembly of a number of claims into a second patent interest not dissimilar to the Bell Company itself, designed to operate competing proprietary services under a parent firm that controlled the rights.

The resultant American Speaking Telephone Company brought together three sets of patents that promised to challenge Bell across a broad front. From the Harmonic Telegraph Company, set up to market Elisha Gray's rights, came Gray's own cluster of harmonic telegraph patents and a pending application that asserted full priority in the invention of the telephone. The Gold and Stock Telegraph Company, which had come under the control of Western Union in 1871, assigned rights acquired from Amos Dolbear, a physicist at Tufts University in Boston who also claimed to have preempted Bell's work. Finally, Western Union contributed the pending claims of Thomas Edison, who proposed a new and far more effective kind of transmitter.[61] Gold and Stock operated as the licensee of American Speaking Telephone in distributing telephones made under these claims, and it was Gold and Stock agents that Bell franchisees confronted in the field from early 1878.

As they embarked on a competitive struggle that lasted until the end of 1879, the two opposing telephone companies had both made intellectual property their organizing asset. At the Bell concern, however, the need for resources to develop the business meant that the company could not long persist as the commercial expression of the four patent owners. Consequently, the early corporate history of the enterprise was strongly shaped by decisions about whether to admit outside capitalists to a place alongside the original patentees. Competition with Western Union immediately exposed the limitations of reliance on the private resources of the two principal backers. Gardiner Hubbard's personal finances were subject to periodic crises, and Thomas Sanders's ability to sustain the company from his own pocket became increasingly strained as his investment passed $100,000 without a return.[62] When Hubbard, managing the company in his capacity as trustee of the patents, began to overstretch its commitments, the discrepancy in spending between the two men proved the spur to changes in the governance of the enterprise.

Largely at Sanders's urging, Bell Telephone began a phased process of reorganizing itself and drawing in outside investors. Hubbard fought a rearguard action against this course, which came at the expense of his own position in the enterprise. At first Hubbard forced a compromise that attached external life support to the company rather than reinvigorating it from within. An injection of outside investment from Boston capitalists in February 1878 was contained by the creation of a separate New England Telephone Company, charged with developing service in that region. Only a few months later, financial need forced Hubbard to admit the New England investors to the original Bell Company, which was refinanced and legally incorporated with a further diminution of the trustee's role.

Although the patentees still held nearly three-quarters of the reorganized company's stock, the new investors received equal representation on the board, and a nucleus of professional management began to supplant entrepreneurial control. Administrative duties now fell on Oscar Madden, late of the Domestic Sewing Machine Company, who took charge of agency affairs, and Theodore Vail, an energetic official of the Railway Mail Service whom Hubbard had recruited as general manager. In 1879 the old Patent Association leadership was formally subordinated to the financiers, and the New England Company and Bell Company were consolidated as the National Bell Telephone Company under a new president, the railroad financier William H. Forbes.[63] Although National Bell, like all the previous incarnations of the company, depended on control of the patents, the ability of the patentees to wield an autonomous interest within the enterprise was at an end.

Competition between Bell and American Speaking Telephone initially remained outside the courtroom, precisely because patents were fundamental to the companies. A potential clash of conflicting patents was an unstable state of affairs. Resolution through the courts might instantly place one party at the mercy of the other, or block both from offering an effective service—both outcomes that the companies would not risk lightly. The obvious alternative was a preemptive settlement, but the uncertainties of the patent situation made that option hard to evaluate. Each side in the telephone struggle held a portfolio of patents on different aspects of voice-communication technology, but the priority and scope—and hence the value—of their respective rights remained indeterminate until they had been matched against each other in the courts. None of the parties necessarily had an interest in testing their intellectual property to destruction. In the end, they did not. The first brief period of telephone competition in America ended with combination, a pooling of the fundamental patents, and the restoration of single-firm ascendancy.

For almost two years, Bell and Western Union licensees (the latter working under the Gold and Stock Telegraph Company) raced to establish telephone exchanges throughout urban America, while the parent companies maneuvered around various versions of a settlement. Negotiations reached an advanced stage in early 1878 and early 1879, but agreement could not be reached. The balance of advantage between the two sides was hard for the principals to calculate, either in terms of competitive position in the field or on the strengths of the respective patents. By May 1879 Bell had perhaps three times as many telephones in service, but Gold and Stock's operations were growing faster.[64] The Bell Company held out strongly in its New England base, but lost its grip on a number of major cities including Chicago and New York. In areas of direct competition—eventually some twenty cities—the Gold and Stock agents could offer lower rates, forcing Bell to come to the aid of its licensees by remitting royalties and in some cases providing emergency finance.[65] More than the struggle for leadership in the cities, the financial weakness of the parent company in Boston threatened to decide the outcome of competition in short order. Full-scale price war was avoided, and for much of 1879 the sides operated a shaky truce while talks progressed. Nevertheless, the need to prop up licensees drained Bell resources and propelled the series of reorganizations that kept the company from collapse in 1878–1879.

Absent the financial meltdown of the Bell enterprise, the potential for a summary victory by either side rested with the patents. An apparent deadlock arose between Bell's basic claim to the telephone and Western Union's ownership of critical working improvements. Both sides pondered long over their legal advice. Hubbard and his colleagues remained confident in Bell's controlling potential, but Western Union had its own redoubt to fall back on. The telegraph company owned rights to Thomas Edison's carbon transmitter, which it employed in place of Bell's magneto instrument (used by Bell as both a transmitter and a receiver). When placed in combination with a Bell-type receiver, the Edison transmitter outstripped the Bell telephone in power and clarity of sound and in range of signal. The improvement in quality, which did much to make the telephone commercially viable, set a standard that the Bell Company needed to match.

Only a pair of timely acquisitions enabled Bell to keep pace on the transmitter front. The first was a caveat filed by Emile Berliner mere weeks before Edison's patent application, also describing a variable-pressure contact transmitter. Tipped off by their lawyer Edward Dickerson, the Bell interests purchased Berliner's rights for $25,000, hired Berliner, and secured an interference proceeding in the Patent Office between his patent application and Edison's.[66] Having bought time by tying up Western Union's application (for what ended up being years) and preventing infringement suits under an Edison patent, the Bell Company then pressed into service its second acquisition, the Blake transmitter.[67] Francis Blake's carbon instrument eliminated the quality gap, but fell within the scope of Edison's pending patent. By 1879 the telephones competing in the field represented clashing rights: Western Union infringed Bell's receiver and his claim to telephony in general, while the Bell Company imitated Edison's transmitter.

Fighting these claims to their legal conclusion seemed to invite a Solomonic judgment under which each party would be able to prevent the other from offering a viable service. In September 1878 the Bell Company did institute a lawsuit for infringement of the basic patents against Gold and Stock and its agent Peter Dowd—a case usually referred to as the "Dowd suit," although Western Union was the real target and took full control of the defense. The two sides began the laborious process of deposing inventors and preparing expert testimony, yet continued to seek an alternative outcome, with negotiations revolving around how the respective patent holdings could be valued. In 1878 and 1879 Bell and Western Union periodically approached agreement on a plan to form a consolidated telephone company. Early discussions, in which the telegraph company sought a majority stake, failed to satisfy Bell investors and executives. Later talks envisaged a new company comprising equal shares for the two sides, with a further chunk of capital to be allotted to one side or the other after arbitration. This formula attracted sufficient support for a $5 million scheme to be drawn up, only for Western Union to withdraw abruptly when its lawyers warned that arbitration might publicize flaws in the patents.[68]

Counter to the consistent format of these proposals, the eventual settlement did not take the form of a consolidated telephone enterprise but of Western Union's withdrawal from the field. Circumstances and the priorities of the telegraph company had changed. The key development was Western Union's reaction to the threat posed by the financier Jay Gould, then mounting a renewed attack on the telegraph business. Not only did Gould establish a rival telegraph system to depress Western Union stock values and open the way for a hostile takeover, but he also sought to gain an attacking foothold in the telephone business, taking a direct stake in some Bell agencies. This move raised the possibility that Gould could gain a competitive edge over Western Union by using his telephone assets as local feeders for his long-distance telegraph network.[69] Western Union chose to cover its flank by expediting the compromise with Bell, on terms that would keep telegraph and telephone service clearly separate. On November 10, 1879, the two companies signed an agreement settling the Dowd case and partitioning the telecommunications sector.[70] In its broad outlines the agreement followed the structure of a compact reached two months earlier between the leading Bell and Western Union licensees in the South, which had ended competition in seven southern states.[71] On the national scale, Western Union withdrew from the telephone business in return for a large royalty and a number of restrictions (later proven ineffective) on telephone encroachment into the long-distance transmission of business messages.

Instead of the pooling of rights taking place through the reconstitution of ownership in a merged company, Western Union's telephone patents were effectively transferred to Bell. All the rights of Gray, Edison, and the others relating to speech transmission passed to the National Bell Company under exclusive license: eighty-four pending and granted patents in all.[72] Western Union in return received a 20 percent share of telephone rental revenues for the lifetime of the patents. The telegraph company's only equity stake was a minority shareholding in a few important urban exchanges: New York, Philadelphia, Pittsburgh, Detroit, and San Francisco. Elsewhere, Bell licensees were left to negotiate terms for the takeover of non-Bell facilities on a case-by-case basis.

The November 10 agreement also opened the way to a final reorganization of the patent-holding company. Investors trumpeted their approval of the end to competition: Bell shares sold at $977.50 in mid-November, almost double the price attained in September.[73] National Bell's leaders took advantage of the high tide in their affairs to expand the company's capitalization, obtaining a special act of the Massachusetts legislature to incorporate the American Bell Telephone Company at $10 million. At the same time, the act placed on a clear legal footing American Bell's power to own stock in other corporations.[74] National Bell had started to take an ownership stake in a few licensee companies in order to support them against Western Union competition. Now the road lay open for American Bell to build on its patent rights.

CHAPTER 3

The Telephone Cases

IF PATENT LAWS WERE NOT SO OPAQUE to historians, the *Telephone Cases* would loom much larger in the history of the U.S. Supreme Court than they presently do.[1] Consider the following headline features. In 1888, the Court upheld the most valuable intellectual property of the nineteenth century, Alexander Graham Bell's fundamental patent for the telephone, by a 4–3 vote. The opinion of the Court was both legally and commercially momentous: legally, as a landmark ruling in the law of patent scope; commercially, because it sustained the monopoly of the American Bell Telephone Company, already a "hundred-million-dollar" corporation, whose prominence in American business life would only increase in later years.[2] A sense of high drama gripped the Court's deliberations. Charges and countercharges of corruption and malfeasance swirled around the case, eventually touching President Cleveland's cabinet and the Supreme Court itself. Finally, the decision was monumental in two further senses. Printed with the arguments of counsel, it filled an entire volume of the *U.S. Reports,* the only case ever to do so. It also stood as a headstone to the "Waite Court," whose chief justice, Morrison R. Waite, took ill after completing the opinion and died four days later. Waite's eulogizers uniformly agreed, in the words of Justice Samuel Miller, that the chief justice's final act closed "one of the most important causes decided while he was on the bench"³—an assessment that ranked the *Telephone Cases* alongside such canonical Supreme Court decisions as the *Civil Rights Cases* or the anti-monopoly *Granger Cases*.⁴

Hard though it may be to believe today, there was a time when a patent dispute could take on such status. To be sure, it needed more than a single case; the Supreme Court hearing was the culmination of a decade's worth of lawsuits fought across the United States. During these battles, the Bell patent question moved beyond the immediate circle of telephone inventors and promoters to become a problem of national import. By the time it came before the Supreme Court, the Bell patent featured in the public life of the country: in journalism, in finance, even in politics. The apparently factual question of who had invented the telephone became a repository for unspoken normative questions, including "Who should control telephone service?" and "Whom should the patent law benefit?" Bell's patent was neither the first nor the last to entangle law, politics, and the pursuit of great fortunes. The historical challenge of such cases lies in working out why the legal results emerged as they did. The devil is in both the details and the big picture: to explain great patent cases of the nineteenth century, we must look to the point where the intricacies of patent doctrine met the wider forces of political economy.

As a legal matter, the Bell Company's achievement of patent control over the telephone rested on two contentions: that Alexander Graham Bell was the first person to develop a practical means of transmitting speech by electricity, and that the method described in his specification governed all subsequent devices for doing so. Establishing and defending these credentials under American law required a substantial exercise in judicial investigation and reasoning, with an eventual body of proof that included a dozen leading judicial opinions and an immense factual record documenting the gestation of telephone technology. At the same time, challengers collectively built up an equally massive body of counterevidence and counterargument with which to overturn the American Bell Company's monopoly. In the face of this onslaught, the proprietary status of the telephone could have slipped, even fallen, in any one of a slew of courts from Boston to New Orleans, and on a whole range of points of fact or law. The fate of Bell's patent thus depended in large part on strategic choices made by lawyers in pursuit of their arguments. More fundamentally, though, it depended on how the courts understood the nature of invention itself.

Technological change is a cumulative process. Inventors typically advance the work of earlier experimenters, improve existing devices, and analyze problems within frameworks of scientific understanding constructed by others. As historians of technology have noted, the resulting trajectory of change can be viewed in different ways. At one end of the spectrum is a vision of incremental and collective effort, leading to new discoveries for which no one person or episode can legitimately take credit. At the opposite end lies a story of sudden leaps forward, discrete moments of breakthrough when groping research crystallizes into an identifiably new machine or method.[5] The difference between sudden breakthroughs and gradual, continuous innovation is analogous to the competing views of biological evolution backhandedly nicknamed "evolution by jerks" and "evolution by creeps." Needless to say, either view of invention has profound implications for the question of what may be patented and by whom.

Descriptions of technical change can, of course, contain both punctuation and continuity. One influential account of technological history suggests that progress typically consists of a few "macroinventions"—radical new technologies such as the steam engine or telegraph that make possible a whole field of innovation or economic activity—and a second, more numerous, order of "microinventions" that follow up with new applications or efficiency improvements.[6] Though undoubtedly more realistic than views that either deny or glorify inventive leaps, even this model has its complications. Fundamental discoveries do not always have immediate impact on society; in many cases the subsequent minor tweak that created a viable product has had far more impact than the original conceptual breakthrough.[7] To some extent, then, the act of marking off innovation into separate inventions always involves drawing artificial lines on a continuous process.

Judges in nineteenth-century patent cases faced precisely this exercise as a highly practical matter. First, in order to allocate rights to the rightful claimant, the courts had to identify discrete moments of invention and attribute them to particular individuals. Under American law, a patent could only legitimately belong to the literal first inventor.

Consequently, whereas other countries' courts had only to discover the first to file a patent, the American method entailed a significantly greater investigation into the history of the invention. Second, judges had to establish the proper scope of each patentee's rights. In order to be meaningful, a patent had to protect against another's use of the machine with only minor changes, say of color or size. But what was "the invention" to be protected? To what extent should a patent cover variations of method, or even improvements?

The first place to look for an answer was in the claims of the patent. Over the course of the nineteenth century, the patent law had developed a formal process by which the inventor was supposed to indicate the subject matter covered by the patent. The 1836 Patent Act included the requirement that a patentee "particularly specify and point out the part, improvement, or combination, which he claims as his own invention or discovery."[8] To satisfy this requirement, applicants appended a short abstract-style statement to the end of the longer specification, typically summarizing the invention and claiming it "substantially as described" or "substantially as set forth" in the long-form description. Interpreting the claim and thus the scope of the patent meant deriving the gist or essence of the invention from the combination of specification and claims. Later in the century, the courts and Congress began to require a more precise and binding delineation of the patent's scope.[9] During the 1870s, claim interpretation shifted, haltingly, from the older "sign post" approach (where the claims plus the specification pointed out the essence of the invention) toward a "fence post" approach, where the wording of the claim alone was meant to mark out the metes and bounds of precisely what was protected.[10] Through the claim, patent law gradually translated the material thing invented (the specific machine or apparatus) into a more abstract "invention" (the protectable idea behind the new thing) and ultimately into words on paper: the text of the claim.[11]

Even during and after the shift to the "fence post" method, though, claim interpretation was never entirely limited to the words on paper. A patent could be infringed by "equivalents" that were substantially identical, even if they fell outside the literal terms of the inventor's claim.[12] And most fundamentally, courts looked beyond the patent document to the originality of the invention described. Not every patented invention was equal; some embodied wholly new technologies, whereas others merely refined existing techniques. These differences translated into different legal outcomes: the narrow patent covering only one particular form of a device, and the pioneer patent, awarded to a radically new technological departure, which granted broader rights over subsequent development. This double standard suffused the law of patent scope. As the Yale professor William Townsend put it, "The first inquiry [in an infringement case] is whether the patent is a primary one; that is, for a pioneer invention—the first embodiment of a means for the accomplishment of the general result. . . . In the case of a primary patent greater liberality is shown in construing its claims so as to protect it against equivalents."[13] Conversely, "if the advance towards the thing desired is gradual, and proceeds step by step, so that no one can claim the complete whole, then each is entitled only to the specific form of device which he produces."[14] Making these determinations embroiled the courts in an examination of the exact nature of the patented invention and the state of the art at the time of patenting.

Powerful practical reasons argued for this kind of differentiation. Courts in the later nineteenth century faced a massive expansion in patenting: the number of patents issued each year rose from under 1,000 in 1850 to over 4,000 in 1860 and over 12,000 by 1870. The cumulative effect of escalating patent grants meant that around 20,000 U.S. patents were in force in 1860, but around 80,000 in 1870 and 180,000 in 1880.[15] Add in the rising volume of litigation resulting from this growth, and judges had strong incentives to separate patents of greater and lesser worth.

At the same time, the notion of a pioneer patent was inevitably a cultural as well as a legal construct. Public depictions of the inventor as individual genius, pushing back the frontier of new technology to the benefit of society at large, already had a long history in the United States, reaching back to the early years of the republic. The debates around the first American patent laws in the 1780s and 1790s included vociferous advocacy—often self-interested propaganda by would-be patentees such as James Rumsey and Oliver Evans—promoting the notion of the inventor as a special class of creative person, distinct from the mere artisan or the introducer of new machinery from abroad, and having a particular claim on the fruits of his intellectual labors.[16] In many respects the design of the patent system itself accepted this underlying view of invention. Like the romantic ideal of the literary author in the sphere of copyright (also generally seen as an eighteenth-century creation), the construction of the genius inventor supplied a privileged figure that influenced the rhetoric and the assumptions of the patent law.[17] As the patent system developed, the cultural influence ran both ways: the law "conditioned men's views of invention while at the same time it was altered to conform better to their anticipation."[18]

In the mid- to late nineteenth century, romantic attitudes toward creativity, the lived experience of rapid industrialization, and publicity campaigns by patent-holding entrepreneurs combined to support a robust popular image of the heroic inventor, embodied in the 1860s by "Men of Progress" such as Elias Howe, Cyrus McCormick, Charles Goodyear, and Samuel Morse.[19] These developments were in constant dialog with the patent system and the law. Patent lobbying and patent litigation had a powerful publicity-generating role: it was no coincidence that Howe, McCormick, Goodyear, and Morse had all launched large-scale attempts during the 1840s and 1850s to gain patent control of national markets for their technologies. On a conceptual level, courts fashioned a legal framework that favored the recognition of heroic invention over the vision of cumulative improvement. In employment disputes, for example, where employers tried to claim ownership over patents arising from their employees' work, judges reflexively protected the employees as the party more closely conforming to the individual inventive ideal.[20] Similarly in the law of patent scope, the courts by the 1870s had gravitated toward an increasingly clear acceptance of broad patents in cases of pioneering invention. As Justice Joseph Bradley of the U.S. Supreme Court noted, almost all inventions "being sought by many minds ... developed in different and independent forms"; yet "if one inventor precedes all the rest and strikes out something which includes and underlies all that they produce, he acquires a monopoly and subjects them to tribute."[21] Pioneer patents provided a legal framework to match a cultural trope: a basis, in other words, for asking who invented "the" telephone.

The mere existence of rules favorable to pioneer inventors cannot, however, explain why the courts applied that designation to Bell. As legal scholars have long recognized, court decisions emerge as much from judicial preference and prejudice as from formal doctrine, however rule-bound they may appear on paper.[22] Just as importantly, the adversarial nature of litigation determined which issues came before the courts and how the parties' arguments were presented. Who sued whom, when, and where mattered a great deal.

§

The Bell Company's first decision to litigate was a delicate one. The year 1878 saw Bell reeling under the challenge of Western Union, the country's dominant telegraph company. Western Union's telephone arm had grown rapidly since its formation earlier that year, and had already overtaken Bell-licensed operations in key cities such as New York and Chicago. Bell executives anxiously debated the wisdom of suing such a powerful adversary, and several times came close to a compromise with Western Union. Litigation finally began in Boston in the fall of 1878. The "Dowd case" (named for the nominal defendant, Western Union agent Peter Dowd) saw the debut of many arguments that the telephone trials of the 1880s would repeatedly traverse.

Most importantly, the Bell Company mapped out its strategy for presenting Alexander Graham Bell's patent. To find Bell's patent fundamental and controlling, the courts would have to identify his invention as a major break from the prior art. Bell's specification explicitly stated the case for this discontinuity in the form of the "undulatory current." As discussed in Chapter 2, early attempts to transmit sound electrically in the 1860s and 1870s had used telegraph-like methods that made and broke the electric current, carrying "galvanic music" but falling short of speech. Bell's key insight was that reproducing complex sounds required a constant and fluctuating current rather than an intermittent one. The "undulatory current," as Bell called it, became the basis of his working telephone and the central feature of his first telephone patent. Stating the theory in a patent, however, was a very different exercise from defending it in court.

Just how different became clear on the eve of trial, when Bell found himself engaged in a "regular pitched battle" with his own lawyers.[23] The attorneys in question were no longer Pollok and Bailey, the Washington, D.C., practitioners who had guided Bell's patent drafting and application. Instead, the generals of the litigation campaign were two Boston lawyers of long experience, both renowned in the patent field. Chauncey Smith was the first general counsel of the Bell firm, "an old-fashioned attorney of the Websterian sort, dignified, ponderous, and impressive." Descriptions of Smith inevitably dwelled on his largeness, his heavy build, his "heavy hair cut round and long like an old-fashioned wig," and a belligerent gaze, behind gold-rimmed spectacles, that shifted readily to a benign twinkle.[24] James J. Storrow was the very image of a patent specialist: studious and undemonstrative, the son of a prominent civil engineer and a graduate of Harvard Law School, "never in the slightest degree spectacular," but deeply immersed in the technological detail of his cases.[25] This contrasting pair—the small, soft-spoken, dark-whiskered Storrow and the large, ruddy, dimpled, and Franklinesque Smith—would be joined in the key Bell patent litigation by another leading attorney, Edward N. Dickerson of New York. Dickerson, an inveterate courtroom grandstander, had a career in patent warfare stretching back to the 1850s, when he had represented Charles Goodyear and managed the notoriously venal congressional patent extension campaign of Samuel Colt.[26]

These were the men who made the Bell patent—and did so in a way that Bell himself had not imagined. The advent of litigation made the division of labor between inventor and lawyer abundantly clear. In the early weeks of 1879, nearly three years after the first telephone patent had issued, Bell and his attorneys disagreed vehemently on how to present it in court. Writing to his wife, Mabel, the inventor described his torment at the sight of

> Chauncey Smith and Company ... dissecting my patents for the purpose of deciding what valid claims there may be on which to base an argument tomorrow. Oh! My beautiful patent of 1876! My specification of which I was so proud! They have hacked it to pieces—They have torn it limb from limb—They have plucked out the *heart* of the invention and have thrown it away—They have cast aside as useless all that I thought most valuable. They have subjected my invention to hydraulic pressure from legal minds and have squeezed out *into the gutter*—the very life-blood of the idea. All that remains of my poor specification is—a *little dry dust*—which they

blow in my face as the essence of the speaking Telephone! It is in vain that I protest. "An inventor is the worst judge of his own case"— "A man should never prune his own fruit-trees" and similar sayings— are rammed down my throat—*ad nauseam*—until I am *choked* into submission.[27]

Bell was scarcely exaggerating the disagreement. He insisted on emphasizing the first four claims of the patent, which outlined methods for generating and transmitting the undulatory current. Chauncey Smith had eyes only for the fifth claim, which applied Bell's method specifically to "transmitting vocal or other sounds telegraphically, as herein described, by causing electrical undulations, similar in form to the vibrations of the air accompanying the said vocal or other sound."[28] Bell became physically ill with worry over "Mr. Smith's seeming lack of appreciation of . . . the fundamental underlying feature of the speaking Telephone."[29] But Smith would not be moved. His reasons were in large part tactical: by concentrating exclusively on the transmission of "vocal or other sounds," he staked the patent on Bell's development of a working telephone, rather than allowing Western Union to drag the trial into a scientific debate about forty years of electrical-acoustic experimentation. Smith also had the pioneer-patent doctrine in mind. By focusing on the claim to speech transmission, the lawyer believed, he could emphasize Bell's concrete achievement and sharpen the technological discontinuity that would allow Bell's rights to be considered fundamental.

When the trial began, Smith's strategy reaped immediate rewards. Bell's adversaries, having prepared to wreck the patent by proving that the undulatory current itself was previously known, found that they had misplaced their defenses. Relenting on his earlier protests, Bell crowed that "[t]hey have stated in their Bill of Objections that my invention was old. . . . Now it turns out to their astonishment that the invention they have stated to be old is not the invention we claim they have infringed and they don't know what to do."[30] The first day of hearings ended with Western Union's counsel fleeing the Boston court for consultations with their client in New York. In a way unexpected to both the inventor and his adversaries, Bell's lawyers had staked his patent on the practical fact of speech transmission, broadly encompassed in the fifth claim. The lasting result of Smith's approach was the installation of the fifth claim at the center of Bell's legal strategy.

Western Union, meanwhile, put forward a comprehensive screen of defenses. These ranged from assertions that Bell's invention lacked novelty to allegations that the machine described in his patent would not work.[31] The most important arguments, from the telegraph company's point of view, involved Western Union's own telephone-related patents. Several of these originated with the engineer Elisha Gray, who, like Bell, had entered the field of sound transmission hoping to develop a form of high-capacity telegraph. Unlike Bell, Gray had focused on his telegraph schemes and did not immediately reduce his voice-transmission ideas to practice, instead filing a provisional notice of priority called a "caveat," which reached the Patent Office on the very same day that Bell filed his own application.[32] Whether this striking coincidence represented simultaneous invention by the two men or suspicious practice on Bell's part would become a controversial subject in later litigation. Other Western Union properties included patents assigned by Amos Dolbear, the Tufts University physicist who claimed to have preempted Bell's work, and Thomas Edison, who had developed an improved form of transmitter.[33] Finally, Western Union's briefs cited many other researchers in electrical acoustics, including Philipp Reis, a German scientist who had constructed and publicized versions of a musical "Telephon" between 1861 and 1864.

These arguments never reached a judicial decision. Late in 1879, Western Union abruptly withdrew from telephone service in order to focus on competitive threats elsewhere in the telegraph business, and the Dowd case was settled.[34] The Bell Company received exclusive rights to Western Union's telephone patents and in return ceded its erstwhile competitor a minority stake in telephone operations. Settlement drew a line under some patent questions, but left others open. A consent decree renounced the claims of Gray, Dolbear, and Edison to priority over Bell, removing at a stroke some of the leading threats to Bell's patent. However, many of the counterclaims raised by Western Union remained available to other defendants. In addition, the Bell Company had still received no ruling on the broad scope of its patent.

Consummation of the strategy launched against Western Union was swiftly arranged through a second suit in the same court. *American Bell*

Telephone Company v. Spencer, filed in July 1880, concerned charges of infringement against Albert Spencer, Massachusetts agent of the small Eaton Telephone Company.[35] Judge John Lowell presided, as he had done in the Dowd case, and the complete record of testimony from the earlier trial was entered as evidence. Although the defendants conceded Bell's priority of invention, the Bell Company pressed for a ruling on the fifth claim of his patent.[36]

Judge Lowell's reading of the fifth claim, handed down in June 1881, made *Spencer* a foundation stone of monopoly. Bell had, in the judge's words, "discovered a new art—that of transmitting speech by electricity—and has a right to hold the broadest claim for it which can be permitted in any case."[37] In ruling Bell's invention a "new art," Lowell set the patent firmly in the pioneer category. Bell had discovered the application of the undulatory current to speech transmission; as such, his rights were not confined to any particular machinery used. This position was crucial because telephone technology had already moved well beyond Bell's original invention. New instruments, especially the transmitters pioneered by Edison, Emile Berliner, and Francis Blake, proved indispensable in making the telephone commercially viable. Yet under the *Spencer* decision, all such advances were controlled by Bell's rights.

Finally, Judge Lowell swept aside Reis. No doubt encouraged by the defense's unsuccessful attempts to make a Reis instrument speak in the courtroom, Lowell erected a solid barrier of legal authority between the "false theory" of intermittent current interruption embodied in the German's design and the constant undulatory current employed by Bell. "A century of Reis," stated the opinion, "would never have produced a speaking telephone by mere improvement in construction."[38] The extent of Bell's victory in *Spencer* was not lost on observers. *Scientific American* informed readers that the judgment "virtually confirms to the American Bell Telephone Company *the exclusive right of talking over a wire by electricity.*"[39]

The core group of cases deriving a broad patent from the undulatory current was completed by *American Bell v. Dolbear,* handed down at the beginning of 1883.[40] Like the Dowd and Spencer cases, the Dolbear case took place in front of Judge Lowell in Massachusetts, but now with an extremely valuable piece in play: Justice Horace Gray of the U.S. Supreme Court, sitting with Lowell on circuit duty. Boston scientist Amos Dolbear, whose own claim to priority had effectively vanished when Western Union surrendered his first telephone patent to Bell, sought to distinguish one of his later telephone models from the coverage of Bell's rights on the grounds that this "condenser receiver" acted by static electricity rather than through electromagnetism. Justice Gray answered with a ringing endorsement of Bell's de facto legal monopoly on speech transmission. The decision's key passage held that "[t]he evidence in this case clearly shows that Bell discovered that articulate sounds could be transmitted by undulatory vibrations of electricity, and invented the art or process of transmitting such sounds by means of such vibrations. If that art or process is (as the witnesses called by the defendants say it is) the only way by which speech can be transmitted by electricity, that fact does not lessen the merit of his invention, or the protection which the law will give to it."[41]

Along with *Spencer*, *Dolbear* ensured that the Bell patent of 1876 would continue to control the field in the 1880s, even though the instrument it described had almost immediately become obsolete. *Spencer* had established that the improved transmitters that had made the telephone commercially viable fell under Bell's rights. *Dolbear* dispatched an attempt to circumvent the patent through technical differentiation and also reinforced the court's uncompromising stance with the authority of a Supreme Court justice.

As one might imagine, the breadth of scope awarded in these cases lay at the margin of what the patent law contemplated. On this point the Bell Company benefited from background shifts in legal doctrine. Patentability was limited by the condition that a patent could not cover a principle of nature, only (at maximum) a process derived from such a principle. Protection for a process in this context could include a broad "mode of operation," such as the use of a hot-air blast in a furnace or, as it transpired, the employment of an undulatory current for speech transmission.[42] However, the line between an unpatentable principle and a patentable process that applied the principle proved to be one of the shifting boundaries of American patent law.[43] In the leading case in this area, *O'Reilly v. Morse* (1854), the Supreme Court had struck down the part of Samuel Morse's telegraph patent that appeared to lay claim to any and all uses of electromagnetism for printing intelligible signs, characters, or letters at a distance.[44] During the third quarter of the nineteenth century, judges deployed *O'Reilly* to limit the scope of patents, invoking the decision in order to invalidate broad claims as descriptions of general principles rather than of concrete inventions.[45]

In the late 1870s and early 1880s, the pendulum swung in the other direction. Justice Joseph Bradley engineered the Supreme Court's new position, validating patents for processes independent of specific means in *Cochrane v. Deener* (1876) and *Tilghman v. Proctor* (1880).[46] The *Tilghman* decision was the more explicit redirection of doctrine. In supporting Richard Tilghman's claim to a procedure for manufacturing glycerine, Bradley upheld the patentee's right to "a patent for a process, and not merely for the particular mode of applying and using the process pointed out in the specification."[47] As in Bell's situation, the actual method described in Tilghman's patent was obsolete, but was nevertheless allowed to control the technology currently in use.[48] Justice Bradley was forthright about the correction in judicial course that he proposed, assailing the misconception that *O'Reilly* "was adverse to patents for mere processes."[49] The judges in Bell's cases proved receptive to Bradley's rehabilitation of the process patent. Writing his *Spencer* opinion shortly afterward, Judge Lowell cited *O'Reilly* to vindicate, rather than challenge, Bell's broad claim. Justice Gray in the *Dolbear* decision based his support of Bell's broad rights directly on the holding of his Supreme Court colleague in *Tilghman*.[50]

What should we make of this evidence that a changing law favored the broad patent sought by the Bell Company? In the context of its legal surroundings, it is clear that this development was not part of a general movement toward greater patent breadth or indeed any one-sided doctrinal tendency supporting patentees. On the contrary, a number of contemporary and historical assessments present quite the opposite picture of patent law in this period. *Scientific American* perceived in the mid- to late 1880s a "recent tendency of the courts to destroy patents" and a Supreme Court "much more vigorous in its treatment of patents than were the old school of judges."[51] Some historians have described the late nineteenth century as a period of judicial reaction against the monopolistic potential of patent rights, during which the Supreme Court endorsed an incremental, "eventless" view of most technological improvement in order to pare down the numbers and scope of grants.[52] At first glance, the support that *Tilghman* gave to a class of powerful patents is inconsistent with these accounts.

However, closer examination reveals discrete parts of the patent law moving separately. The decisions cited as evidence of tightening patent policy can be traced to two doctrinal developments. The first is the Supreme Court's stand against the abuse of patent reissues. Reissue of specifications, which in theory allowed patentees to amend honest mistakes in patents already granted, had become associated with attempts to broaden the scope of rights and adapt the text for litigation.[53] During the 1870s both Court and Congress toughened the line against dubious reissues, finally cementing the requirement that reissued patents could be no broader than merited by the original grant.[54]

The second significant doctrinal shift supposed to have worked against patentees is a hardening policy on minor patents. Responding to arguments that an excessive profusion of property rights obstructed industry, the Supreme Court during the 1880s repeatedly asserted a requirement of inventive creativity—what later generations would know as the doctrine of "nonobviousness."[55] This rule, which had appeared in the 1850s but was widely implemented only after the Civil War, distinguished patentable inventions from routine improvements by requiring that the former demonstrate "more ingenuity" than possessed by "an ordinary mechanic acquainted with the business."[56] The Court of the 1880s insisted on this distinction, maintaining in the leading case of *Atlantic Works v. Brady* that "[i]t was never the object of those laws to grant a monopoly for every trifling device, every shadow of a shade of an idea, which would naturally and spontaneously occur to any skilled mechanic or operator in the ordinary progress of manufactures."[57]

If we can detect any consistent pattern in these lines of cases, it is not favor or disfavor for powerful patent rights as such, but a desire to better differentiate between inventions of differing worth. Broad and narrow patents coexisted, with the difference between the two being determined by a clear criterion: the originality of the invention. As judicial policy, this preoccupation with sharpening the analysis of inventive contribution had its merits. In the context of successive expansions in patenting, it addressed concerns about the quality of inventions receiving protection and encouraged judges to sort the wheat from the chaff. This sorting in turn strengthened the basic rationale for the social value of patents and allowed an increasingly pro-property-rights judiciary to stand fully behind the rights of inventors.[58]

Even more importantly, an emphasis on differentiation left the courts with great flexibility to deal with different technological situations. Much depended on judges' fact-specific interpretation of the patents and inventions before them. As a result, while the doctrines of patent scope did not take account of economic monopoly, contestation of powerful patents was possible in plenty of other ways. In particular, much of the resistance to monopolistic patents took the form of arguments that the grant was invalid in the first place.

With the Boston court ruling out any route around Bell's rights, attempts to invalidate the patent came to the fore. Again, activity was channeled by the peculiar configuration of American law. Whereas most countries recognized the first claimant to file a patent, U.S. patents issued only to the "first and true" inventor—a standing invitation, gratefully accepted by Bell's growing number of opponents, to put forward prior inventors "discovered" after the technology had become a going concern.[59]

The telephone litigation is justly remembered for this colorful cast of characters: men such as the mechanic Daniel Drawbaugh of Eberly's Mills, Pennsylvania, self-described as "one of the greatest inventive geniuses of this age," who alleged that he had built telephones in the 1860s and early 1870s.[60] Another claimant, Dr. Sylvanus Cushman, maintained that he had done so in Racine, Wisconsin, in 1851, when his electrical experiments had suddenly enabled him to hear the croaking of frogs in a nearby swamp.[61] A third, the Italian-born machinist Antonio Meucci, claimed to have invented a speaking telegraph while employed as a theater decorator in Havana in 1849 or 1850. Meucci had filed a caveat (notice of invention) with the U.S. Patent Office in 1871, but had not pursued a full application, he explained, because an explosion on the Staten Island ferry had rendered him an invalid.[62] These men and others like them could point to extensive personal histories of electrical experimentation. But their central qualification was prior obscurity, accompanied by pleas of poverty to explain why they had not publicized their discoveries sooner. The Bell Company, anxious to rebut the pretenders' claims, responded by hiring the Pinkerton's detective agency to ferret out discrediting details.[63]

Not all rival telephone companies uncovered their own claimant. Many invoked the work of Philipp Reis, whose scientific reputation enjoyed a remarkable posthumous resurgence in the mid-1880s.[64] The continuing appeal of Reis's invention to Bell's opponents deserves some explanation. At root it was an attack on the hard line the Boston courts had drawn between Bell's undulatory current and all previous acoustic telegraphy. Using a device that, like a telegraph, alternately made and broke an electric current, Reis had transmitted musical tones in the early 1860s. It did not take much—only the weighting of an armature, or the tighter stretching of a diaphragm—for Reis's "Telephon" to operate with the constant and fluctuating current employed by Bell. When thus configured and very carefully handled, the Reis telephone could talk.[65] On this basis, Bell's opponents argued that Reis had invented the telephone (some claiming that he had transmitted actual speech during the 1860s), or, at the very least, that the Boston court's rigid distinction between undulatory-current telephones and make-and-break devices ran ahead of scientific understanding of electricity. The editorials of *Scientific American* provided early and vocal encouragement for Reis's claims, although this enthusiasm almost certainly stemmed from the involvement of at least two of the journal's editorial staff in rival telephone companies.[66]

Though popularly remembered as bids for inventive credit, challenges to Bell's priority were ultimately driven by financial considerations. The number of claimants was, as *Scientific American* remarked, "a faithful index of the value of the prize."[67] The search for potential anticipations worked in the same way as the regular market for inventions, with investors either gathering behind or seeking out promising claims. Antonio Meucci remained obscure until brought to the attention of a Philadelphia syndicate, whose members organized a company backed by the Baltimore & Ohio Telegraph interests. Sylvanus Cushman attracted support from Chicago city councilors and drugstore owners antagonistic to the local Bell operating company.[68] Perhaps the most important connection occurred in 1879, when a Washington patent lawyer named Lysander Hill represented Daniel Drawbaugh and his partner Edgar

Chellis in a patent dispute over a faucet. Shortly afterward these men formed a partnership to promote Drawbaugh's telephone claims, joining forces with businessmen from New York, Washington, and Cincinnati to incorporate the People's Telephone Company at an authorized capital of $5 million.[69]

The People's Company joined a diverse band of telephone enterprises established in defiance of Bell's patent. It is impossible to account for all of them; even the six hundred infringement suits initiated by Bell did not uncover every backwoods exchange and workshop-built telephone. However, the most determined infringers—those who led the legal fight against Bell—adopted a characteristic speculative model. After incorporating with a collection of minor telephone patents and a large paper valuation, these ventures promoted operating companies in multiple states, aiming to profit from the sale of licenses and stock. Thus the New York–based Molecular Telephone Company licensed an offshoot in Cleveland, Ohio, and the Overland Telephone Company promoted subsidiaries in Pennsylvania, New Jersey, and Kentucky.[70] The Pan-Electric Telephone Company, formed in Tennessee, marketed its patents to parties in Missouri, Illinois, Alabama, Texas, and the Washington, D.C., area.[71] Some of these ventures resulted in the construction of actual telephone lines; others remained on paper. However, all shared a common aim: to stave off the inevitable infringement suit from Bell.

The campaign launched in Daniel Drawbaugh's name demonstrates the scale of this effort. In response to a suit filed by Bell in New York, Drawbaugh's lawyers deposed dozens of witnesses to his invention of the telephone, collecting an unprecedented eight thousand pages of testimony evidence over three and a half years.[72] Fortunately for Drawbaugh, financial stamina and eminent legal representation were hallmarks of the People's Telephone Company. The first attorney to argue Drawbaugh's case was George Harding, leader of the Philadelphia patent bar. Later the company retained Senator George Edmunds, the ablest constitutional lawyer in Congress, who had declined President Arthur's offer of a Supreme Court seat and was himself a credible contender for the Republican presidential nomination in 1884.[73] In the interest of partisan balance, Edmunds would be joined by Don Dickinson, a prominent party man on the Democratic side and a confidant of President Cleveland.[74]

By the time the Drawbaugh case came to a hearing in New York, lawsuits across the country waited on the result. Observers eagerly anticipated a clash of legal heavyweights. To counter the prestige that Senator Edmunds brought to Drawbaugh's representation, the Bell Company had retained another powerful Republican, former Senator Roscoe Conkling of New York, who had declined the same Supreme Court seat offered to Edmunds.[75] To the disappointment of the newspapers, however, neither man played much part during two weeks of dense and often technical argument. On the Bell side, James J. Storrow lectured indefatigably on the origins of the telephone before methodically attacking each of Drawbaugh's claims to the invention. Lysander Hill, for Drawbaugh, then embarked on a marathon tour of alleged anticipations of Bell, after which he reviewed at similar length the testimony on Drawbaugh's prior invention and the poverty that had made Drawbaugh unable to exploit his discoveries. Bell attorney Edward Dickerson then entered to pound the railing of the courtroom and to damn Drawbaugh as a professional impostor and "bunko steerer." Conkling lounged indifferently and read the newspaper; Edmunds sat with his chin in his hand.[76] The result of the case did not disturb Bell's monopoly. Judge William Wallace concluded the trial by rejecting Drawbaugh's claims and casting aspersions on his inventive ability. Drawbaugh's own words, noted the judge, revealed "without the aid of extrinsic evidence, the ignorance and vanity of the man, and . . . suggest[ed] also the character of a charlatan."[77]

After Judge Wallace's ruling for Bell, the other dominoes fell quickly, most of them in the same New York court. Wallace issued injunctions against the McDonough group of telephone companies in February 1885, the Molecular Telephone Company in March, and the Overland firm in December. Judges in Pennsylvania, New Jersey, Ohio, Kentucky, Louisiana, Maryland, and Texas followed suit.[78] As the rain of decisions for Bell came down, companies tried unsuccessfully to shelter themselves by alleging new evidence, by abandoning subsidiaries in mid-trial so as not to be bound by the verdict, and by attacking the legitimacy of earlier decisions, claiming that defendants such as Spencer had secretly colluded with Bell.[79]

Throughout the process, the federal courts answered their mandated role as "an indivisible system for ascertaining the rightfulness and the limits of the patent."⁸⁰ This function fell to judges, rather than juries. Patent cases by the later nineteenth century were overwhelmingly tried as cases under the equity jurisdiction of the court, where the judge alone was the fact finder and decision maker, rather than under the federal courts' parallel common law jurisdiction, where a jury was responsible for deciding factual questions of infringement and validity. Patentees and their lawyers explained their choice of equity proceedings in terms of the desire for judicial expertise, but it was surely not lost on them that judges were a safer bet than juries in cases of powerful and controversial patents.⁸¹ As the flourishing interstate trade in patent assignments indicated, judges supplied patent enforcement that was reasonably reliable from one state and jurisdiction to the next.⁸² And as Americans had begun to discover, the federal courts served as a crucial support for the legal and financial operations of large corporations.⁸³ These generalizations held in the telephone cases. Federal circuit judges in Boston, New York, and Philadelphia—the men who held responsibility for the consistent enforcement of patent rights across a whole swath of industrial America—observed a firm rule of respect for one another's rulings. Crucially, the impulse for uniformity extended beyond the major patent forums of the Northeast: the Bell Company was able to win its decisions in the South and Midwest.⁸⁴ Geography, by forcing Bell to bring multiple suits, had provided the telephone patent infringers with some room to maneuver. Even so, it provided no refuge from the nationwide enforcement of Bell's rights.

§

In the mid-1880s the telephone patent question moved beyond the world of financiers, lawyers, and engineers onto a broader public stage. There the reputation of the Bell patent became entangled with two great preoccupations of the Gilded Age: corruption and monopoly. The charges of corruption were various and reached to the highest echelons of government. First, opponents bested in the circuit courts opened a new front by alleging massive fraud in Bell's original patent application. Several different variations on the fraud charge appeared, the most serious being the accusation that Bell's lawyers had stolen the variable-resistance concept from Elisha Gray's caveat with the aid of a corrupt Patent Office official and

The mid-1880s saw the telephone patent battles repeatedly enmeshed in scandal. The image of Uncle Sam shocked and perplexed by telephone skullduggery was a recurring motif. Thomas Nast, "The Telephone Scandal. Hello! Hello!! Hello!!!," *Harper's Weekly,* February 11, 1886, 107.

added it to Bell's application. These charges led to a federal government lawsuit being initiated against Bell.

From there, the scandals only mounted. Further complicating matters, the government's intervention was itself ensnared in allegations of corruption. It quickly became clear that the government suit was a front for the Pan-Electric Telephone Company, one of whose directors was Augustus Garland, then attorney general in President Cleveland's administration. Amid newspaper outcry the president ordered an end to proceedings against Bell, and Congress launched an investigation of Garland's role in the Pan-Electric affair. Finally the remaining branch of government was dragged into discredit by allegations of impropriety on the part of judges who had heard Bell cases, including Justice Gray of the U.S. Supreme Court.

Driving the descent into scandal was a mounting hostility to the Bell Company's monopoly. Political agitation for telephone rate regulation first appeared in the Midwest, cradle of the American antimonopoly tradition and heartland of the Granger movement, an agrarian front that had secured state price control of railroads and other monopoly services in the 1870s. Moves to impose rate limits on the telephone began in Indiana in 1885 and rippled across the state legislatures of Illinois, Ohio, Missouri, Pennsylvania, and New York. At the same time, big-city users began to translate their frustration with the variable quality and high cost of telephone service into pressure on the operating companies and the Bell parent firm. The New York City Board of Trade, no enemy of patent rights, endorsed the government's fraud investigation on the grounds that "the manner in which the public have been treated in this city by the [Bell-affiliated] combined telephone companies is entirely unjustifiable." Even in Massachusetts, the American Bell Company's attempt to increase its statutory capitalization touched off a round of proposals to regulate telephone rates.[85] Under the combined pressure of corruption charges and regulatory challenges, Bell's stock price declined by 33 percent during 1885.[86]

Politicized, newspaper-fueled agitation against the Bell monopoly added fresh challenges to the company's legal situation. As long as the circuit court cases hung in the balance, Bell executives lived in fear of a judge with "any taint of grangerism or any political bee in his bonnet."[87] Interventions by elected officials further complicated the picture. After the initial Garland scandal, the federal government renewed its fraud suit—a legal second front on which the Bell interests would be forced to fight for years. Meanwhile, on the local level, municipal governments occasionally threw support to infringing companies, not always for wholly public-spirited reasons. In Chicago, for example, the city council responded to strong anti-Bell agitation by granting a franchise to the Cushman Telephone and Service Company, in which a number of councillors had a personal interest.[88]

The most important battle of these years, however, involved an audience of just nine men: the justices of the U.S. Supreme Court. The Court elected to consolidate the appeals of Dolbear, Drawbaugh, and the Molecular and Overland telephone companies into a single hearing, thereby promising a definitive ruling on Bell's rights. In the year before argument, which took place in January and February 1887, Bell's lawyers had reason to be confident. Every rival claimant had lost in the lower courts, and several speculative bubbles had been pricked: in one notorious example, the properties of the United States Telephone Manufacturing Company, valued in the company's books at a million dollars, sold at auction for $100.[89] During litigation the Bell patent had already received favorable rulings from Supreme Court Justices Matthews and Blatchford on circuit, in addition to Justice Gray's sweepingly broad construction in *Dolbear*.[90]

Against these advantages for the monopoly, those whose sympathies lay with the challengers found some causes for comfort. Recent decisions by the Supreme Court against apparently monopolistic patent holders—primarily arising from its new reissue doctrine—gave credence to the notion that Bell's rights would be defeated or narrowed.[91] In addition, the Drawbaugh claim had held on to its wealthy backers.[92] Most encouraging of all for Bell's opponents was the arrival of the company's scandals at the door of the Supreme Court itself. In December 1886 the *New York Herald* revealed that numerous relatives of both Judge Lowell and Justice Gray had held Bell stock at the time of the crucial Boston telephone cases. Both judges denied having known of these investments when they made their rulings, but Gray, whose family still held stock, was forced to recuse himself from the Supreme Court's deliberations.[93]

Given the timing of the revelations and the obvious benefits of knocking out *Dolbear*'s author, it is fair to assume that the anti-Bell interests were behind the *Herald*'s scoop. Newspapers were crucial links in the Gilded Age system of financial information and disinformation, and both the Bell Company and its opponents had long engaged in news placement—paid and unpaid—on behalf of their stock prices.[94] Either the anti-Bell telephone companies or the lawyers handling the U.S. government's fraud suit might have launched the scandal. Much of the information revealed by the newspaper had previously been fed to the government's attorneys by John McClay Perkins, a Boston patent lawyer. Perkins had written a series of letters to the attorney general in July 1886, fulminating about Bell corruption and detailing Gray's shareholdings. He would later claim that "the revelations in the *New York Herald* of Dec. 2 1886—all were discovered by me."[95]

Further complicating the web of interests connecting the press and the Bell question was the role of party politics, especially surrounding the government's suit. The scandal over Attorney General Garland's involvement with the Pan-Electric Company had been largely driven by New York newspapers, particularly the *Tribune* (which broke the story), *Sun, World,* and *Evening Post*. Conversely the *New York Times,* which had already developed an anti-Bell stance, set itself against this group and supported the government's case against Bell.[96] Politics probably determined where the newspapers stood on the Bell patent. In a politically fluid period for the New York press, the *Tribune* and *Sun* at least were implacably hostile to President Cleveland's Democratic administration, though the *World* was a Democratic paper; all condemned the corrupt origins of the government suit.[97] On the other side, George Jones, proprietor of the Cleveland-endorsing *New York Times,* was an established foe of the Bell Company. The *Times* never hesitated to condemn American Bell as an "odious monopoly" and its main patent as a "fraudulent issue."[98] While the Pan-Electric affair rumbled on, Jones opined that outrage against the administration had been orchestrated from behind the scenes by the Bell monopoly: "The Pan-Electric scandal, involving certain public men at Washington, is a small matter in comparison with the Bell Telephone scandal, involving certain newspaper editors in the city of New York."[99]

All of this muckraking ensured that fraud accusations took a large role in arguments before the Supreme Court. Lysander Hill of the People's Telephone Company provided the most spectacular moment of the hearings when he advanced a new version of the corruption charge against Bell. Hill claimed that the Bell patent application held on file at the Patent Office was a forgery—a "clean copy" smuggled in to conceal telltale signs that Bell's original application included material stolen from Elisha Gray. As evidence, Hill produced a document that had appeared in the record of the Dowd litigation between Bell and Western Union in 1878–1879. This copy, Hill said, showed an original version of Bell's patent application covered with hasty-looking penciled amendments. The penciled additions included numerous important features of the issued patent, including the variable-resistance method that Gray had described in his caveat. Here was proof, Hill contended, that Bell's lawyers had gained access to Gray's caveat in the Patent Office and had altered the Bell patent application after it was submitted.[100]

This argument proved disastrous. Bell counsel James J. Storrow countered that Hill's evidence was in fact a different document entirely: a copy of an earlier specification drawn up for an English patent application. The pencil markings were, Storrow asserted, his own notes on Bell's subsequent changes, somehow mistakenly left among the records of the Dowd case.[101] Even Bell opponents present at the court agreed that Hill's allegations were "paper-thin" and that Storrow had torn them apart, while reporters noted soberly that the incident "was generally regarded as a point scored by the Bell company."[102] The justices had "no hesitation in rejecting" the fraud claim in their eventual decision.[103]

Other than the allegations of corruption, arguments in the Supreme Court hearing rehearsed those made in the circuit courts. The accumulated briefs and testimony submitted in evidence now totaled twenty-two volumes and fifteen thousand pages, another record. Grosvenor Lowrey, the veteran litigator of electrical patents, was the leading spokesman for Philipp Reis among the attorneys present. His task was to dispose of the technological discontinuity upon which Bell's broad claim depended. To this end he sought to dispel Bell's "pioneer" status, giving a careful account of electrical-acoustic history and adding summaries such as a "Resume of Material Facts known to Physicists in 1861," in order to demonstrate that Bell's discovery was far from a new art in 1876.[104]

If Hill offered the justices corruption and Lowrey gave them science, the lawyer-politicians George Edmunds and Don Dickinson gave the court Daniel Drawbaugh's life story.[105] This part of the case was all about credibility. Judge Wallace in New York had used the weaknesses in Drawbaugh's account, especially his dubious pleas of perpetual poverty, to dismiss his claim as "*falsus in uno, falsus in omnibus*" (false in one thing, therefore false in all). In response, Edmunds and Dickinson worked to shore up the trustworthiness of the inventor and his witnesses. Storrow, replying to them, attacked Drawbaugh's failure to profit from his supposed invention: "If [the courts] cannot read the telephone in the events of his life, they will not accept it from his deposition."[106] Of all the arguments mounted against Bell during the hearing, Drawbaugh's claim seemed to attract most interest from the bench, a development that translated immediately into movements in the value of Drawbaugh Telephone Company stock. On the other hand, the monopolist's strong share prices showed that the balance of confidence remained in favor of Bell.[107]

When the Court issued its verdict over a year later, both assessments proved correct: the contest came down to a straight fight between Bell and Drawbaugh, and Bell won. With Justice Gray recused from the case and Justice Woods having since died, the remaining justices divided four to three. The conclusion of the *Telephone Cases* (as the judgment is still known) cemented for posterity the image of a closely fought and closely decided litigation. Without going into great counterfactual detail, it is reasonable to say that the history of the telephone would have been quite different had a single vote on the Court gone the other way. What, then, led the justices to rule as they did?

Ostensibly the outcome hinged on the credibility of the rival inventors' claims to priority. The Court's majority opinion, delivered by Chief Justice Waite, hewed closely to the Bell Company's account of Alexander Graham Bell's invention. Meanwhile the minority, led by Justice Bradley, noted that Drawbaugh's claim seemed "so overwhelming, with regard both to the number and character of the witnesses, that it cannot be overcome."[108] The difference between the court's two wings, Bradley suggested, lay in their willingness to accept the reputational advantage acquired by Alexander Graham Bell during ten years of relentless legal campaigning. It was, Bradley observed, "perfectly natural for the world to take the part of the man who has already achieved eminence. No patriotic Briton could believe that anybody but Watt could produce an improvement in the steam engine. This principle of human nature may well explain the relative feeling towards Bell and Drawbaugh in reference to the invention of the telephone."[109] Bradley, for his own part, spoke up for the inventiveness of the "plain mechanic" and the possibility of serendipitous, untheorized, and subsequently underexploited invention.

Questions of invention, then, appear to have decided the case. However, courts—perhaps especially the U.S. Supreme Court during the Reconstruction era—have been known to dispose of cases in a way that minimizes or disguises the real issues at stake.[110] The great mystery of the *Telephone Cases* is whether the justices responded to other pressures: in particular, to the heated politics of the Bell Company's monopoly. The composition of the minority suggests that some did. Of all the justices, dissenters Field and Harlan were the ones least engaged with patent questions: Field wrote nine majority patent opinions in his thirty-four years on the court; Harlan wrote four in thirty-three years. Furthermore, both were prolific dissenters in general.[111] Given their wordless affirmation of Bradley's dissent, it is quite possible that they voted in part to chasten the Court's majority and the Bell monopoly. This is not to say that their actions were entirely capricious; Harlan, for one, conducted a substantive discussion with Bradley about the Drawbaugh witnesses.[112]

Nor can Bradley's vote and opinion be lightly explained away. Justice Bradley was both one of the court's leading authorities on patents and one of its chief skeptics of industrial monopoly.[113] On several occasions he had condemned attempts to stretch patents into illegitimate monopolies and warned against perverting the patent laws into "instruments of great injustice and oppression."[114] On the other hand, he had done much to fashion the recent law of patent scope, including its emphasis on broad patents for pioneer inventions.[115] Indeed, Bell's chief legal arguments rested on opinions Bradley had written. If Bradley wished to rein in the Bell monopoly, then his best option was to do so using the factual record. His painstaking trawl through the Drawbaugh evidence, revealed by the notebook he kept during the hearings, may be best viewed in this light.[116] Otherwise it was out of character: Bradley typically applied the skepticism toward obscure claimants that his colleagues brought to bear on Drawbaugh.[117]

Bradley's dilemma was, in many ways, that of the patent system more generally. The great project of nineteenth-century patent jurisprudence had been the development of rules for matching the scope of protection to the inventive contribution made by the patentee. In the eyes of treatise writers, the law of the 1880s and 1890s finally formed "a harmonious, symmetrical, scientific system" for translating specified inventions into property rights.[118] Yet this conception of the patent as a reward to individual genius was at best complicated, and at worst dysfunctional, in the face of patents deployed as instruments of corporate power. The justices who upheld Bell's patent resolved this tension in the predominant mode of the period: they looked strictly to the rights of the inventor and took no notice of the market consequences of the patent. For Joseph Bradley, it is fair to surmise, this approach may not have sufficed.

If Bradley's dissent buried tensions of monopoly within questions of invention, it was only fitting. In the telephone cases and in American patent law more generally, the two issues were thoroughly entwined. Under a system that privileged pioneer inventions with one hand while leaving them vulnerable to unknown prior claimants with the other, any valuable patent could expect to be assailed on priority grounds. So long as the "monopoly question" remained a prominent part of American political economy, patent law had the potential to attract legal and economic conflict.

§

The courts framed the question of "who invented the telephone?" in two ways. They asked, and answered, the "who?" They also established the subsequently unexamined assumption that there was such a thing as "*the* telephone." Both legally and historically speaking, the most notable aspect of the telephone cases is not that the courts accepted Bell's priority of invention, but that they granted vast scope to his patent—endorsing, in doing so, a unitary theory of telephone technology and its origin. Much of the credit for this phenomenon must go to Chauncey Smith, James J. Storrow, and the other architects of Bell's legal campaign. These men seized a chance offered by the preexisting legal environment: a strong judicial regard for pioneer patents, which translated into a willingness to grant broad protection in certain cases. Furthermore, Bell's lawyers fought off the hazards that awaited such a claim in the nineteenth-century courts, ranging from eruptions of obscure prior inventors to the unacknowledged pressures of antimonopoly sentiment on judges' decisions.

In many ways, the telephone patent battles followed a classic pattern. As a Morse, Howe, or Goodyear could testify, any broad grant that promised to control a new technology was inevitably assailed at law by the patentee's competitors or customers. The Bell Company, however, faced additional challenges. Federal government intervention against the Bell patent was something new: a product of strengthened post–Civil

War government and a growing resistance to patent monopoly prompted and sustained by the rising tide of populism. The political economy of patent law was changing, producing both legal confusion and legal innovation. Nowhere were these responses better represented than in the government's fraud suits against Bell.

CHAPTER 4

The United States versus Bell

THE BELL MONOPOLY TESTED THE LIMITS of patent law. Nowhere was this clearer than in the U.S. government's attack on the Bell patent—a remarkable intervention beset by scandal. At the height of the telephone litigation, the federal government launched a lawsuit charging that Bell had obtained his patent by fraud. The resulting outcry shook President Cleveland's administration and spread charges and countercharges of corruption across the national press. Yet the Bell case was only the latest in a line of attempts to draw the federal government into action against monopolistic patentees. It would become an important test of the government's power to cancel patents.

In the twenty-first century, national governments have an established, if periodically controversial, role in correcting for apparent failures to serve the social ends of the patent system. Provisions such as compulsory licensing of pharmaceutical patents, special conditions for government use of proprietary technology, and forced divestiture as part of antitrust enforcement are all employed to some degree across the developed world. In the late nineteenth century, however, the facilities for states to further diffusion of new technologies at the expense of patentees' rights were few. Most countries employed a "working requirement" that provided for cancellation of a patent if the technology it protected was not manufactured locally within a certain time period—commonly two years.[1] The major exceptions were Britain and the United States, which, even more than other nations, allowed patentees freedom to supply or to withhold their patented technology as they saw fit.

The economic returns to an individual patent *could* be regulated after issue in Britain and America, but only in one direction: both countries allowed the extension of grants where the patentee could show that he had been prevented from receiving an adequate reward. Conversely, patentees who restricted the supply of their invention, or who exercised monopoly control over the market for their product, faced no sanction. The United States, with its generally staunch pro-patentee regime, eschewed almost all methods of government interference with patent property. Not only was a working requirement consistently rejected by Congress, but for much of the nineteenth century the federal government even lacked the power to revoke a demonstrably invalid patent.[2] Antitrust law, which did not arise on the federal level until late in the lifetime of Bell's patent, in any case did nothing in those years to limit the use of patents in creating or sustaining monopolies.

When the U.S. government stirred into action against Bell, then, it was not because of any formal role in regulating the market effects of patents. Instead the government was ostensibly acting to seek revocation of an allegedly fraudulent grant. This effort was a thoroughly *ad hoc* mobilization of government power against the telephone's patented status, with the United States acting as an interested opponent of the monopoly rather than as an impartial regulator.

The federal government would assail the Bell Company's patent position on two occasions. The first government suit alleged fraud in the issue of Bell's 1876 patent, based on a number of shady details that came to light during the Bell litigation. The suit itself had deeply murky origins, but would become something more than its corrupt beginnings: a test case that probed the boundaries of federal power to cancel patents and led both the executive and the courts into unknown legal territory. A second fraud case, involving the Bell Company's Berliner patent, revisited the question of patentees' duty to the public and provided an important marker in the Supreme Court's turn-of-the-century attitude toward corporate patent power.

§

It is entirely possible that Alexander Graham Bell received his first telephone patent partly with the aid of official misconduct. The circumstances of February and March 1876, when the U.S. Patent Office received and approved his application, include several instances that give pause. The dramatic near-simultaneous submission of Bell's patent and Elisha Gray's caveat strongly suggests that Bell's lawyers had advance intelligence of Gray's plans. Within the Patent Office, everything seemed to fall right for Bell: the withdrawal of the interference hearing initially declared with Gray, the timely amendments to his pending application that steered Bell clear of another Gray interference, and the rapid issue of Bell's grant. The financial hold that Bell's lawyer, Marcellus Bailey, had over the patent examiner Zenas Fisk Wilber completes the picture—and leaves open the possibility of still more corrupt acts, including the outright theft of crucial ideas from Gray's design.[3]

All of these charges and more were leveled against Bell during the litigation of the 1880s. For our purposes, the actual merits of the accusations directed at Bell and the Patent Office are less important than the legal logic that brought them to the fore. In fact the same was true for Bell's accusers, who went through several versions of their allegations in pursuit of a charge that would stick. In their attempts to secure a government suit, Bell's opponents at first dressed up as "fraud" the simple charge that his patent lacked novelty, arguing that the inventor had falsely claimed his discovery as new. Subsequently they added claims of misconduct in the Patent Office, pointing to Bell's amendments as evidence of illicit access to confidential information on other applicants' work.[4] The alleged wrongdoing deepened in the following year, when patent examiner Wilber signed an affidavit admitting to having favored Bell over Gray in closing the interference and to having revealed the contents of Gray's caveat.[5] Charges of fraud reached their fullest extent in the Supreme Court's telephone appeals, when opposing counsel Lysander Hill alleged that Bell had not only enjoyed illegal access to Gray's caveat but had copied its crucial variable-resistance method into his own application.[6]

The legal attraction of claiming fraud was twofold: it provided Bell's challengers with new grounds on which to conduct litigation once their other arguments were spent, and it promised to lend a veneer of government authority to their case. Both were necessary to prop up the infringers' speculative project, which was well served by any protraction of the anti-Bell campaign. In addition, the group that instigated the government case almost certainly hoped that Bell would buy them off. As a minority congressional report later noted, "[W]hat they wanted of the Government was a weapon which they could thrust into that [Bell] company to bleed it."[7] Motive and opportunity came together in 1885, when two of the anti-Bell companies collaborated in seeking a government suit.

One of them, the Pan-Electric Telephone Company, had turned hungry eyes to Washington, D.C., ever since its formation in Tennessee in 1883. The Pan-Electric was the brainchild of J. Webb Rogers, a Tennessean whose son Harry had secured a number of telephone patents. While other infringers had built their power bases in New York, Pennsylvania, and New Jersey, the Pan-Electric had disbursed directorships and paper capital of $5 million among southern Democratic senators and officeholders, encamped subsidiaries in Maryland and the District of Columbia, and lobbied for a contract to install telephones in the Capitol.[8] The company's prospects were greatly strengthened by Democratic victory in the presidential election of November 1884, after which several Pan-Electric directors took high positions in the new administration. When the anti-Bell legal campaign turned into an undignified rout in the summer of 1885, the Pan-Electric interests moved to exploit their political assets.

In late June and early July, the Pan-Electric and its fellow southern enterprise, the National Improved Telephone Company of Louisiana, abandoned the infringement cases they were defending on behalf of subsidiaries in Pennsylvania and shortly afterward agreed to petition jointly for a government proceeding against Bell. The telephone companies undertook to conduct and to pay for the case themselves, employing only the government's nominal sponsorship. Attorney General Augustus Garland, a Pan-Electric shareholder, formally declined to authorize a suit, only for the solicitor general to do so in his absence. Pan-Electric lawyers duly began proceedings in Memphis, Tennessee, political base of

The Pan-Electric Affair became a major embarrassment for President Cleveland's administration and for Attorney General Augustus Garland in particular. In this political cartoon, a serpentine Pan-Electric telephone has Garland's ear. Thomas Nast, "That Garland Has Slipped," *Harper's Weekly,* February 11, 1886, cover illustration.

one of the company's most powerful directors. When politically hostile elements of the New York press cried scandal, President Cleveland ordered Garland to halt the case.

The government's entanglement up to this point represented an unvarnished advancement of private interests. Historical accounts have accordingly tended to bracket the episode as one of corrupt adventurism, as did contemporaries.[9] Yet it would be overly narrow to see the telephone fraud suit simply in terms of Gilded Age "influence" and Pan-Electric scheming. The Bell case was only the most visible of a number of attempts at the time to draw the federal government into fiercely contested patent actions.

§

The catalyst for these efforts was an 1871 case, *Mowry v. Whitney,* in which the Supreme Court considered an attempt to secure the cancellation of an allegedly fraudulent patent.[10] Justice Samuel Miller's opinion in *Mowry* analyzed the English law of revocation by scire facias: an action that allowed interested parties to bring suit in the name of the monarch to cancel patents that had been inappropriately or fraudulently granted. Some twenty scire facias actions had taken place in England before the procedure fell into disuse in the 1850s, the most famous instance being the 1785 suit to revoke Richard Arkwright's spinning-machine patent on the grounds of an insufficient specification.[11] Miller concluded that the appropriate American equivalent should take the form of a suit brought in the name of the government. Private parties alone could not mount a case for revocation. In Miller's formulation, "The general public is left to the protection of the government and its officers."[12]

The years that followed saw a steady trickle of attempts to put Miller's ruling—a judicial prescription without any statutory mandate or institutional precedent—into practice. Petitions arrived at the Department of Justice seeking the attorney general's imprimatur for privately conducted suits, prompting no small amount of administrative and judicial confusion. The practice of referring petitions to the commissioner of patents stymied several applications; unsurprisingly, successive commissioners proved unwilling to confirm charges of fraud in their own department. The few cases authorized to proceed received a mixed hearing in the courts. In Rhode Island, Illinois, and Ohio, judges refused to accept government revocation suits in the absence of a statute.[13] But in New York, the leading patent jurisdiction, Circuit Judge John Wallace allowed one such case to go forward as the logical extension of *Mowry v. Whitney.*[14]

Two characteristics of the various government fraud proceedings are particularly relevant to the Bell telephone suit. First, the lack of clarity in the law frequently led to rather vague charges. In the absence of a definition of "fraud," some petitioners offered only general arguments of invalidity—essentially warming over unsuccessful arguments from earlier trials. In the case of *United States v. Colgate*, for example, Judge Wallace found "no allegations in the bill charging fraud or false suggestion on the part of the applicant in his application for a patent. At most, the allegations show that there was no novelty in the invention, and inferentially that he, knowing the prior state of the art, which was public knowledge, must have known there was no novelty."[15] Arguments of this kind found little favor with the courts, and for good reason. Judge Henry Blodgett of Illinois pointed out that the practice, if allowed to spread, would "transfer nearly all litigation on patents, except mere questions of fact as to infringement, to the office of the attorney general."[16] The Pan-Electric interests began their campaign by making similar charges, although they subsequently escalated their accusations of active wrongdoing. In all likelihood Bell's opponents knew they had to come up with stronger versions of the fraud claim, especially once it became clear that the courts took a dim view of revocation suits based on lack of novelty alone.

Second, government suits were disproportionately sought during high-stakes, sometimes highly political, patent struggles. One of the earliest post-*Mowry* fraud suits, relating to Eben Norton Horsford's valuable patent on calcium phosphate baking soda, arose out of a sprawling legal fight already taking place in the courts of New York, New Jersey, South Carolina, Georgia, and Rhode Island.[17] Another concerned an audacious attempt to exert control over wood-planing technology, using an old 1847 application revived and patented in 1873. The owners of this belated grant (the Woodbury patent), which purported to control all modern planing machines, immediately began to demand royalties from hundreds of lumber firms. Lumbermen responded by forming a protective association to defend against infringement suits and at the same time sought a revocation proceeding from the government. In this case the government proceedings came to nothing, probably because the commissioner of patents reported no evidence of fraud, and traditional litigation brought an end to the Woodbury patent a few years later.[18]

Two more government cases in the 1880s, however, repeated the Woodbury suit's combination of mass litigation and organized response. In the oilfields of Western Pennsylvania, E. A. L. Roberts patented and developed a nitroglycerine torpedo for blasting wells that quickly became the industry standard. In establishing his dominance of the business, Roberts filed over two thousand infringement suits and dispatched spies across oil country to identify the "moonlighters" who defied his patent.[19] Hard-fought litigation during the 1870s ended with the Roberts torpedo concern firmly in control, and in 1882 independent oil producers petitioned the attorney general for a suit in the name of the United States.[20] At the same time another huge patent fight, this time centered on Iowa and Illinois, raged over barbed wire. The Washburn and Moen Company sought to control the manufacture of this essential tool for prairie farming through its fundamental Glidden patent, suing rivals and threatening farmers across the Midwest with lawsuits. Iowa farmers organized and counterattacked: the state legislature requested a revocation suit from the U.S. attorney general while the Iowa congressional delegation pressed for reform of the patent laws at the national level.[21]

The political turn of these two suits was no coincidence. The architect of the barbed wire plan was the same man who had petitioned for the Roberts torpedo suit on behalf of the citizens of Western Pennsylvania and who had played a similar role in the matter of the Woodbury planing patent: General Benjamin F. Butler, a Massachusetts lawyer and radical politician of national renown. The general's taste for controversy and his political ambitions—he ran unsuccessfully for president as the People's Party candidate in 1884—drew him to such populist legal causes. Gadfly though Butler was, to the Iowans fighting the barbed wire monopoly, he was also a champion who could be expected to take their case for a mere $250.[22]

In each of his three major attempts at a government suit, Butler used political momentum to obtain the attorney general's authorization for action in the name of the United States. Yet none of the cases resulted in a full court hearing. The Woodbury prosecution stalled, and the torpedo suit, initiated only months before the Roberts patent expired, apparently deterred the Roberts interests from pursuing further prosecutions and was settled. The barbed wire suit was also settled after Washburn and Moen cut their prices by 75 percent.[23] All three patents eventually went to the U.S. Supreme Court as a result of traditional patent litigation. Nevertheless, by bringing the possibility of federal intervention to the center of some of the bitterest patent battles of the 1870s and 1880s, Butler did more than anyone else to carve out a place for government revocation proceedings in the wake of *Mowry v. Whitney.*

The rural activism of Butler's clients may seem a long way from the urban settings in which the early telephone industry flourished. But patent campaigns in the 1870s and 1880s were just as important in rural and extractive economies as they were in the industrial sector. And the antipatent sentiment generated by these cases and others like them spilled over to affect the national political economy of intellectual property. Western agrarian hostility to patents was a substantial force, aroused by some inflammatory attempts to demand royalties for basic rural articles.[24] Farmers channeled their resistance to this rent-seeking through the same political institutions that they had successfully deployed to seek railroad regulation: granges and protective associations campaigned on the legislative front as well as in court, by pressing for reform or even abolition of the federal patent law.[25] In their political challenge to the patent system, the farmers found unlikely allies in the railroads, heavy consumers of invention who were trying to reduce their own exposure to infringement litigation. Further support came from a number of "Mugwumps"—good-government reformers from the northeastern states, who defended intellectual property rights on principle but concurred with many of the railroads' proposals on efficiency grounds.[26]

The combined influence of these groups made weakening of American intellectual property law a distinct possibility in the 1870s and 1880s, and participants in the patent-dependent high-technology industries well knew it. Bell counsel Chauncey Smith recalled being assured by a Massachusetts congressman that a large number of his fellow federal legislators stood ready to repeal the patent law at any moment.[27] As late as 1888 the National Electric Light Association withdrew its plan to petition Congress for a commission of inquiry into patent reform, having been warned by patent lawyers that to do so might invite an attempt to sweep away the entire system.[28]

For a great patent monopoly, then, the 1880s were years during which any branch of government could present an avenue of attack. A window of opportunity had opened in case law for attacking grants through the federal executive, while the security of patent rights was only a few votes away from being compromised by Congress. Three bills introduced between 1880 and 1884 would have *required* the attorney general to seek revocation of fraudulent or mistaken grants. The last of these passed the House in 1884, only to fail in the Senate.[29]

The Pan-Electric suit, for all its reliance on the Garland connection, was launched against this broader background. Bound up in the well-known tale of "influence"—itself a feature of nineteenth-century law that should be considered endemic—are at least two important lessons about America's patent regime. First, that it was possible to make an end run around the courts, using other branches of the federal government either to extend or to assault a patent. The 1880s may have seen a flurry of anti-patent activity, but such tactics had a long history: opponents of Charles Goodyear's controlling rubber patent had tried to pass a law for the repeal of fraudulent grants in 1850.[30] The second lesson was that the rules of the game could be changed even as the protagonists played it. Where the law remained unclear or unstable, as it was in the question of patent cancellation, parties had every incentive to try to nudge the system in their favor through statutory reform, courtroom argument, or both. In the telephone fraud suit, the courtroom eventually emerged as the primary battleground. Because Congress stood firm against attempts to weaken the patent law, the government case against Bell became a test case that would resolve the federal power of cancellation one way or the other.

It would do so, however, without the direct involvement of the telephone companies that had initiated the suit. From an instrument of private litigants, the fraud case changed into something else: a genuine government suit, under the control of the Department of Justice. This reconstitution began with President Cleveland's hasty intervention over the Garland scandal. Officially, the reason for halting the first politically disastrous proceedings in Memphis was that the standard practice of referral to the Interior Department had not been followed. Quickly redressing this omission, the telephone companies repeated their petition and secured a hearing before the secretary of the interior. Here the infringing companies again made their case, now seconded by voices from the Bell Company's increasingly unquiet big-city customer base.[31] Interior Secretary L. Q. C. Lamar's report, delivered in January 1886, must have pleased Bell critics at large more than it did the Pan-Electric and National Improved interests in particular. On the one hand, Lamar recommended that the suit go forward and set out a broad basis for review of Bell's patent that included not only fraud but also general invalidity. On the other hand, Lamar called for a suit "in the name of and wholly by the Government, not on the relation or for the benefit of all or any of the petitioners."[32] The Pan-Electric and National Improved companies thus lost the ability to settle with Bell on preferential grounds, which was probably their objective all along.[33]

Following its new mandate, the Department of Justice brought together what Edward Dickerson called its "Bell Telephone Annex."[34] Of five attorneys initially retained, Senator Allen G. Thurman of Ohio was the obligatory political heavyweight, and Grosvenor P. Lowrey the leading patent specialist. Thurman, a distinguished septuagenarian Democrat, faded from the latter stages of the case through a combination of ill health and the toil of running for vice president on President Cleveland's reelection ticket. Lowrey, meanwhile, simultaneously represented the Molecular Telephone Company in the Bell infringement case before the Supreme Court, bringing to the government's case a keen awareness of the intertwined fortunes of the different telephone suits.

The government lawyers necessarily plotted their strategy with reference to the private infringement litigation because, given a broad remit to challenge the validity of Bell's patent, their arguments actually differed little from those of the private defendants. Although Lowrey personally professed to believe the charge of fraud against Bell, he had little faith that it could be proven and did not detect any serious effort being made to obtain clinching evidence on this point.[35] The evidence of fraud rested largely on the testimony of former Patent Office examiner Zenas Fisk Wilber, who claimed to have facilitated Bell's theft of the variable-resistance idea from Elisha Gray's caveat. Wilber's self-confessed dishonesty and history of contradictory affidavits made this avenue an unpromising one for the government suit. The former examiner's fate is one of the sadder aspects of the Bell litigation. Increasingly succumbing to his alcoholism, Wilber was taken to Denver by his Secret Service escort in the hope that he would sober up. He drank himself to death there in August 1889.[36]

Instead, the meat of the government case consisted of impeaching the Bell patent's originality in favor of Philipp Reis. Lowrey hitched his case to the Reis bandwagon created by earlier litigants (including his own clients), while incorporating the lessons learned in the circuit courts. Most importantly, the new suit would concentrate on witness evidence that Reis had transmitted speech, rather than relying on scientific nuance—or, as other government lawyers put it, the "great quantity of sworn speculative essays on electricity dumped into the telephone litigation"—to sway the judges.[37] Spirits in the government camp ran high when an agent reported new Reis evidence from Germany. The great question in 1886, however, was how to coordinate these arguments with the Supreme Court's impending hearing of the infringement cases. Lowrey argued for a full commitment to the Supreme Court case, to the extent of stipulating the government's new evidence into the record for use by Bell's challengers. After all, he reasoned, if the Court ruled on the validity of Bell's rights, "we cannot escape being compromised in an equal degree, whether we have endeavored to aid the Appellants or not."[38] Even before any judge had ruled on the government case, a decision of sorts beckoned at Washington.

In the meantime, to the extent that the government attorneys could make their own fate, they sought to do so in their choice of jurisdiction. Understandably determined to avoid Bell's home state of Massachusetts, Lowrey favored New Jersey and the scientific orbit of his pro-Reis allies at Princeton.[39] Eventually Columbus, Ohio, emerged as the chosen destination, and suit was filed there in March 1886. The reasons for this decision are obscure, save that most jurisdictions in the East had already ruled for Bell in an infringement case. Ohio, or more particularly Circuit Judge Baxter, was suggested to the Department of Justice by a former licensee of the Molecular Telephone Company.[40] Critics of the suit later alleged that Baxter was a notorious "patent smasher," although—possibly unknown to the lawyers of the Telephone Annex—he was also one of the few judges to have heard and dismissed a patent cancellation case brought in the name of the United States.[41] In any event, Baxter's death before trial removed any expected advantage from that quarter and in the process left the government with a jurisdiction problem. Serving suit on American Bell in Ohio by virtue of the business done there by its licensees had been a risky maneuver and ultimately caused the circuit court in Columbus to dismiss proceedings.[42]

The failure of the government's case to obtain a full hearing in 1886 meant that the issues involved were addressed piecemeal, in a succession of proceedings over the following years. Defeat in Ohio left the members of the Telephone Annex with no option but to bring suit in Massachusetts, where Judges LeBaron Colt and Thomas Nelson heard arguments in the summer of 1887. American Bell moved for a hearing on the government's power to cancel patents, and secured another dismissal of the case. Judge Colt's opinion, against which the government immediately appealed, surveyed the conflicting case law and came down on the side of those rulings that had insisted on the need for a specific statutory power of cancellation. Congress, after all, had failed to provide one despite several bills introduced for that purpose.[43] Shortly afterward, the substantive parts of the case received a decision of sorts when the Supreme Court decided the infringement cases for Bell early in 1888. Just as Lowrey had feared, the defeat of Reis and affirmation of Bell's fifth claim effectively stripped the government case of its chief arguments.

Against this background, the government's own Supreme Court appeal on the power to cancel patents, argued half a year later, seemed somewhat moot. Nevertheless, the Bell Company mounted a vigorous case in opposition, during which James J. Storrow urged the court to consider the essentially private nature of intellectual property contests. Storrow argued that all parts of the patent system, adjudication included, were "based for their operation and motive power on personal interests. The great work of making inventions, perfecting the machines and pushing them into public use rests solely on private enterprise and initiative. The lesser work of litigation may well be trusted to the same forces."[44] Even in England, Storrow pointed out, scire facias had long ago given way to infringement suits between private parties as the recognized means of negotiating contested rights in invention. In the same vein, Storrow questioned the government's standing to bring a fraud suit where it had no direct pecuniary interest: unlike in the case of land patents, which granted federally owned land, the government gave up nothing in creating exclusive rights in invention.

The Court, however, did not acquiesce in this vision of a free and automatically self-regulating patent system. To the argument that the federal government had no stake in cases of fraud, Justice Miller replied for the Court that the United States bore an "obligation to protect the public from the monopoly of the patent which was procured by fraud."[45] Miller had argued before for a government role in correcting mistaken and fraudulent patent issues, not least when he had authored *Mowry v. Whitney* some seventeen years earlier.[46] Finally he was able to lead his colleagues in creating the explicit power of cancellation that Congress had declined to furnish. With the government's authority confirmed, the Bell case was sent back to the Massachusetts circuit court for a hearing on the merits.

It may have consoled the government's attorneys to have made the law of the land, but their case returned to Boston in a weak state and went downhill rapidly. Grosvenor Lowrey had sadly to admit that "[n]o judge has ever seen . . . what I think I see" in Philipp Reis's invention, which had become a casualty of the Supreme Court's first telephone ruling along with Dolbear, Gray, and the major charge of fraud.[47] On the other hand, Daniel Drawbaugh's near-success gave grounds on which to refocus the suit, especially given changes in the Supreme Court's membership. Chief Justice Waite's death removed the deciding vote against Drawbaugh. Meanwhile the separate appointment to the Court of former Interior Secretary L. Q. C. Lamar, who had authorized the government's fraud case while in office, raised the possibility of a new majority adverse to Bell.[48] The rump Drawbaugh interests took a close interest in proceedings, which now largely concerned reexamination of their witnesses, though there is no evidence that they provided concrete support.[49]

The government case, however, would never see a final judicial decision. Collection of testimony dragged on as the Bell Telephone Annex dwindled to a single attorney, Charles S. Whitman. Under his direction, the government's deposition of witnesses lasted two years and two months and concluded only months before the expiration of Bell's first patent; the Bell Company then took testimony until 1895.[50] Whitman's death in 1896 effectively ended the suit.

§

No sooner had the first government case against Bell become bogged down than a second took its place. This new affair concerned Emile Berliner's patent for the microphone, a grant which managed the impressive feat of matching Alexander Graham Bell's for notoriety almost as soon as it was issued. Berliner's patent potentially controlled all transmitters that employed variable pressure between contacts—that is, the practical form of transmitter used since the late 1870s. Yet the Berliner grant appeared only in November 1891, fully fourteen years after application, and threatened to extend the Bell Company's monopoly until 1908. The suspicion that inevitably surrounded such an irregularly issued patent of such enormous value to American Bell propelled the federal government back into court against the telephone company.

Berliner's grant was a classic example of what is now commonly known as a "sleeping" or "submarine" patent. Both terms refer to a patent delayed—whether deliberately or not—for a long period during the application process, before "surfacing" on issue to take a mature industry by surprise.[51] Like extended patents, grants of this type are able to circumvent one of the basic features of the patent system: that much of the term of a broad fundamental patent may run while the financial returns on the technology are unripe. Depending on the rate of uptake, an invention may not be profitable until well after its pioneers (including the holders of any foundational patents) have invested in installing, developing, or marketing the new product or process. A submarine patent, by contrast, combines the broad rights of the pioneer patent with the far greater returns of an established industry. Even more than a broad patent issued in timely fashion, a sleeping grant may control generations of the technology unimagined at the time of the original invention.

Understandably, given these properties, submarine patents have supported some of the most hotly contested patent monopolies. The Woodbury planing-machine patent, subject of mass litigation and another government suit, provides a good example: the original application failed in 1849, only to be revived and issued in 1873 after the 1870 Patent Act had provided a grace period for reopening old applications.[52] The 1870s also saw controversy over Charles Page's patent for the electrical induction coil, issued by special act of Congress thirty years after Page's 1838 invention, then deployed by Western Union against telegraph and signal companies.[53] A more calculated delay characterized George Selden's automobile patent of 1895. Selden, a patent lawyer, kept his 1879 application in the Patent Office with a series of continuation motions, periodically amending it to keep up with new developments. His assignees then unleashed the grant on the automobile industry after 1900, reaping royalties for a decade before Henry Ford successfully defied the Selden rights.[54] A century later these issues continued to resound in the patent law thanks to the billion dollars in royalties earned during the 1990s by Jerome Lemelson, whose patents for bar codes and machine vision were based on applications originally made in the 1950s.[55]

As these examples suggest, submarine patents exhibit varying degrees of premeditation. Berliner's patent was not a forgotten property opportunistically revived, a "Rembrandt in the attic" like the Woodbury and Page inventions; it was a live application whose significance for the telephone industry was always clear.[56] American Bell claimed that the examination process had been under the control of the Patent Office throughout, but the company's opponents naturally suspected a deliberate delay orchestrated somehow by the monopolist.

Both could agree that the fourteen-year wait had broadly comprised three stages. From Berliner's caveat filing in April 1877 until June 1882, the application faced a number of obstacles including interference proceedings, the sudden rejection of Berliner's application in December 1881, and its subsequent reinstatement on appeal to the commissioner. This first phase did not attract any later accusations of wrongdoing.[57] The second period, lasting until the Supreme Court's telephone decision in March 1888, amounted to a long wait for resolution of the Drawbaugh claims. Because Drawbaugh professed to have invented the microphone (along with every other basic telephone component) ten years before Berliner, the latter was put on notice of interference.[58] None of the parties involved, however, saw much point in setting up a Patent Office hearing on Drawbaugh's claims while the courts were investigating the matter on such a grand scale. Consequently, the Bell and Drawbaugh companies and the Patent Office came to a "tacit understanding" that all parties would wait until the matter was judicially decided. Whether this hiatus was procedurally required (the Bell position) or unlawful and collusive (the government's charge) became a central question of later litigation.[59] Finally, the third period saw two further hurdles for Berliner's application. In May 1888 the Patent Office examiner again rejected the specification, necessitating another reversal by the commissioner. The grant then remained on hold while a last proceeding flushed the remnants of the Drawbaugh application out of the Office—again, a period that the government would later point to as one of unnecessary delay.⁶⁰ Drawbaugh received his dismissal in October 1891, and the following month Berliner's patent came into force.

Adverse press reaction was immediate, but did not immediately trigger government scrutiny of the Berliner grant. Instead, as in the Pan-Electric affair, the catalyst for government intervention was a personal link between one of the Bell Company's rivals and a senior federal law officer: in this case electrical manufacturer Milo G. Kellogg and the U.S. solicitor general, Charles Aldrich. Kellogg's involvement with the telephone industry was deep and long-standing. A prodigious inventor in the field, he held 152 granted or pending telephone patents in 1892. As superintendent of the Western Electric Manufacturing Company in the late 1870s, he had supplied both sides of the Bell-Western Union competition in Chicago.⁶¹ When Western Electric became American Bell's manufacturing arm in 1882, Kellogg expanded his interest in telephone service by taking a large personal stake in some of Bell's regional licensees. His relationship with the parent company was fractious. In 1887 he brought suit against American Bell as a minority shareholder in the Bell-controlled Great Southern Telephone Company, claiming that Bell milked the company with instrument rentals at the expense of dividends. Bell agreed to pay a dividend for five years, but on the expiry of the arrangement Kellogg's grievance resurfaced.⁶²

It was from this semidetached position within the Bell-aligned family of companies that Kellogg determined to act against the Berliner patent, claiming that he did so in order to free Bell licensees from burdensome royalty payments.⁶³ In November 1892 he sent a brief critiquing the Berliner grant to Charles Aldrich, his lawyer during the shareholder action against Bell and now solicitor general in the Harrison administration. The personal connection between the two men proved crucial in launching a government suit, not least over the strong objections of Commissioner of Patents William Simonds, who had issued the Berliner patent.⁶⁴ Attorney General William Miller appointed Robert S. Taylor, a patent lawyer and judge from Aldrich's former home town of Fort Wayne, Indiana, to report on the matter. Taylor found grounds for a government suit and subsequently took charge of preparing it. His groundwork was enough to ensure that the case survived the change of administration in

March 1893. In winning over the incoming attorney general, Taylor no doubt benefited from the resolutions of support that began to arrive from antimonopoly leagues and from business associations dissatisfied with the state of telephone service.[65]

Government counsel knew from the start that the direct case for fraud in the Patent Office was weak. Taylor looked instead to the Berliner patent's flaws in drafting and to the potentially invalidating effect of an overlapping grant issued to Berliner in 1880. These concerns had already proved substantial enough to provoke two rejections by the patent examiners. By contrast, the charge of willful delay rested on the innuendo that, given the Bell Company's power and the enormous value of a fifteen-year extension to the telephone patent monopoly, the process must have been within the firm's control.[66] This claim had some corroboration: a previous commissioner of patents had reported to Congress in 1888 that "powerful, rich, and influential parties" exhibited a "desire to prolong the issue of their patents"; newspaper coverage suggested that the commissioner may have meant the Bell Company specifically.[67] However, no direct evidence of deliberate delay existed, whereas American Bell could produce correspondence showing that it had requested issue at various points. At worst, the Bell Company argued, it had been overly passive in awaiting a decision from the Patent Office. The legal case for "laches" (unreasonable delay) therefore depended on holding Bell to a uniquely high standard of diligence and justifying that standard with reference to the great public importance of the telephone monopoly.

A proposition firmly rejected by the courts in the Bell infringement cases thus came to the fore in the Berliner suit, namely that the courts should consider a patent's monopolistic implications when testing it against the basic requirements of the law.[68] The Bell Company's brief before the Circuit Court of Massachusetts called this a "Kansas populist notion."[69] It became an altogether more plausible idea when Judge George Carpenter ruled for the government. Carpenter, a Rhode Island district judge sitting by special designation in the Massachusetts court, held that since "[t]he result of any delay which might take place in the issue of the Berliner patent would evidently be to continue so much longer the practical monopoly of the art of electrical transmission of articulate speech . . . the duty of the respondent corporation was to use the greatest degree of diligence in prosecuting the application to an early issue."[70] Judged by this enhanced standard, the Bell Company had "intentionally acquiesced" in the Patent Office's sloth "for the purpose of delaying the issue of the patent."[71]

Taylor rejoiced at this "high and wholesome" doctrine, and embraced the delay issue as the central plank of the government's argument on appeal.[72] The circuit court of appeals, however, unequivocally rejected the idea that Berliner's patent should be treated differently because it perpetuated a monopoly. American Bell's motives, its market position, and the value of the microphone patent were irrelevant, the court declared. To vary the standard of review by such criteria would be "to deny that the laws are equal, and would furnish a standard for the determination of the rights of patentees too fickle and imaginative to form a proper basis for the use of a court of law."[73]

The government's case suffered conclusive rejection when Berliner's patent came before the Supreme Court in 1897. Justice David Brewer, writing for a six-to-one majority, found no evidence of fraud and affirmed the decision of the court of appeal. In doing so, Brewer went beyond denying that a monopoly patent held special status. He articulated a forceful vision of the patent as the "absolute property" of the inventor: "Counsel seem to argue that one who has made an invention and thereupon applies for a patent therefor, occupies, as it were, the position of a quasi trustee for the public; that he is under a sort of moral obligation to see that the public acquires the right to the free use of that invention as soon as is conveniently possible. We dissent entirely from the thought thus urged."[74] In this view, no government power could encroach on the patentee's prerogative to do as he pleased with his rights: "He may withhold the knowledge of it from the public, and he may insist upon all the advantages and benefits which the statute promises to him who discloses to the public his invention. . . . No representative of the public is at liberty to negotiate with him for a new and independent contract as to the terms and conditions upon which he will give up his invention."[75]

This view of the patentee's sovereign rights was not particular to the Berliner case, but represented part of a significant movement in American patent law in the 1890s and 1900s. During these years, an influential line of decisions made the inviolability of patent property into a fetish, to the point where the patentee's rights began to overspill the boundaries of the invention itself and protect the owner's ability to attach "tying" conditions to licensing and use. Judges reasoned that, because a patentee had the right to withhold his invention from use entirely, he necessarily had the right to license it on any conditions he wished.[76] To give one notable example from the case law of the time, the manufacturer of a patented mimeograph machine was allowed to require that purchasers use only the same company's nonpatented ink and paper, on pain of a lawsuit for "contributory infringement."[77] Restrictive practices of this type waned in the 1910s, when the Clayton Act of 1914 brought restrictive patent licenses within the purview of the antitrust law and the Supreme Court reversed its approval of tying requirements. Nevertheless, from the mid-1890s to mid-1910s the principle of absolute patent rights was ascendant. As one circuit judge in Chicago proposed, "Within his domain, the patentee is czar. The people must take the invention on the terms he dictates or let it alone for 17 years. This is a necessity from the nature of the grant. Cries of restraint of trade and impairment of the freedom of sales are unavailing, because for the promotion of the useful arts the constitution and statutes authorize this very monopoly."[78] Justice Brewer's opinion in the Berliner suit was not only representative of this judicial posture but a seminal statement of it, consistently noted in the leading cases of patent-monopoly jurisprudence for the next half century.[79]

As with the first installment of *United States v. American Bell,* however, the Berliner suit made more difference to the law of patents than it did to the market structure of telephony. American Bell's monopoly did not long survive the expiry of Bell's two basic grants in 1893 and 1894. Gradually, then in accelerating numbers, non-Bell ("independent") telephone companies began to provide service. The government's case offered early independent companies a degree of legal cover against threatened suits under the Berliner patent and became a rallying point for the midwestern electrical manufacturers who led the promotion of non-Bell exchanges.[80] But Berliner's victory against the government in 1897 neither surprised nor unduly worried the independents.[81] The Supreme Court's ruling had addressed only the government's charges of fraud, leaving the patent's substantive flaws open to attack in any subsequent infringement case. In addition, the Court's recent case law included a precedent suggesting that Berliner's patent would fall because of a similar grant issued to the same inventor in 1880. This decision was so unfavorable to Berliner that government lawyers expected a Bell attempt to reverse the holding by legislation.[82] Predictions that the former monopolist would be unable to enforce another controlling patent eventually proved correct, when the circuit court in Boston held Berliner's patent invalid in 1901. The court of appeals sustained the patent two years later, but construed the invention so narrowly as to render Berliner's rights irrelevant.[83]

By finally hauling down one of the Bell Company's keystone patents, private litigation had succeeded where a dozen years of government efforts had failed. In retrospect, of course, the line between "public" and "private" in the government telephone suits was far from clear. Even after the federal executive had ceased to act as a facade for the Pan-Electric interests, its campaigns against American Bell were essentially a second front for the parties already in court against Bell's monopoly. The cases thus recall Stephen Skowronek's description of American government as a porous "state of courts and parties": in this instance, with the executive branch serving as a channel for private interests rather than acting on independent institutional commitments or developing a permanent bureaucratic remit.[84] To look at it another way (because recent American historiography has moved away from the idea of a weak state and a "party period"), the monopoly telephone patents were inherently political objects. Private struggles over their distributional consequences were predisposed to become public and governmental affairs, just as American political economy ultimately made all issues of technology and development into matters of public contestation.[85] In either interpretation, public and private were overlapping fields, rather than opposing fiefdoms.

§

The government suits against Bell present a historical puzzle. Were they entirely anomalous and capricious actions, or an indication of new institutional commitments to regulating patent power? The evidence suggests something in between. The telephone presented a special case for reasons that ranged from the uncommon strength of the patent monopoly to the political connections of would-be market entrants. But the government actions also involved institutional innovations and forced officials to justify their positions in both legal and political terms. Once we strip away the garish overlay of scandal and corruption that covered the government fraud suits, this story tells us something about the position of the state in the landscape of intellectual property.

Although the saga of the United States versus American Bell embodied the political sensitivity of patent monopoly and even established the law of patent revocation, it did not carve out a lasting role for the state in policing patent power. Two main factors prevented the government from being embroiled in further attempts to check overmighty patents. First, by requiring that an action for fraud be under genuine government control and that it demonstrate actual misconduct rather than general invalidity, the Supreme Court halted the freewheeling use of the "fraud" charge that had developed in the 1870s and early 1880s. In formalizing the power to cancel a fraudulent patent, the Court ended the legal uncertainty that had allowed General Butler and others to forge the government fraud suit into an antimonopoly weapon.

Second, the pressures that produced high-profile, politicized suits aimed at invalidating a great controlling patent simply became less relevant as the single-great-patent model was superseded in the early twentieth century. Although pioneer patents continued to appear, industrial corporations came to depend less on individual grants than on thickets of patents acquired by purchase or by continuous research. Likewise, corporations found pooling and restrictive licensing arrangements to be both more durable and more readily created by managerial fiat than broad pioneer patents. As the nature of patent monopoly shifted, so too did those aspects of it that most concerned policy makers. By the beginning of the twentieth century, government engagement with patent power had become an aspect of the antitrust effort, focused on collusive interfirm arrangements and restrictive licenses.

Ultimately, protest against powerful patents, as expressed in the Bell fraud suits, reflected the strength rather than the weakness of patent rights. Even in Congress the supporters of the patent system were strong enough to prevent the passage of hostile legislation, and the federal judiciary adopted an increasingly unyielding pro-patent posture toward the end of the century. As a result the telephone cases failed to disturb the fundamental acceptability of broad patent monopolies under American law. When the Berliner suit attempted to impose heightened standards on the owner of a broad grant, the appeals court and Supreme Court demurred, refusing to link the patent's validity to its importance. The Supreme Court then categorically rejected the notion that the public had any claim on a patentee's rights during his allotted term of ownership.

The last decade of the nineteenth century and the first decade of the twentieth amounted to a high point of patent rights' freedom from public interference. Up to and during the 1880s, state and federal courts dismissed the idea that the sale and use of patented articles (as distinct from the rights themselves) were somehow free from regulation. Tests of this principle included suits by Bell-affiliated telephone operating companies in a half dozen states, in which the companies argued unsuccessfully that their exclusive patent licenses should excuse them from some of the requirements of state common-carrier laws.[86] From the late 1890s, however, the absolutist view of patent rights that had inflected the Berliner decision led the federal judiciary to change course and except patent arrangements from some otherwise applicable law. Most notably, the Supreme Court initially blocked the application of antitrust law to patent pools and held that even a price-fixing pool was a legitimate exercise of the patentee's "absolute freedom in the use or sale of rights."[87] Only in the mid-1910s, when the Court's majority shifted against patent pooling and leasing restrictions, did government finally establish its role in regulating market control through patents.[88]

CHAPTER 5

Atlantic Crossings

FROM THE START, the telephone story was an international one. Even as Alexander Graham Bell prepared his American patent application, he had begun moves to protect and exploit his invention abroad. This task involved more than just submitting foreign patent applications. Bell and his backers scrambled to form partnerships with distant investors and to strike agreements with foreign governments, all while promoting Bell's inventive claim and reputation abroad. Not all the challenges were local in nature. The Bell interests soon found themselves locked in competition with other American inventors, particularly Thomas Edison, in securing foreign patents and establishing overseas companies. The prize over which these inventors fought most bitterly was Great Britain, the world's great imperial and financial hub. The ensuing Bell-Edison battle in Britain saw the inventors wrestle with unreliable local agents, ungovernable capitalists, and a government suspicious of competitors to the state-owned telegraph system. In the process, they revealed the challenges of promoting new technology on a global scale.

Among the pathways of international patenting, the route between Britain and the United States was particularly well travelled. In 1877, 217 American patents (1.6 percent of the total) were issued to Britons, who made up over one-third of all foreign patentees.[1] Patenting in Britain had long been more internationalized: in the decade after the 1852 Patent Act, one-fifth of applications came from foreigners. During the period 1867–1869, Americans submitted 826 applications to the British Patent Office (7.2 percent of the total), second only to French inventors among the foreign contingent. By the time the Patent Office began systematically to collect nationality data in 1884, Americans accounted for 1,181 applications (6.9 percent of the total) and were the best-represented foreign group.[2]

This is not to say that overseas patenting was straightforward. Depending on the nature of the inventor's contacts in the destination country, various types of intermediary might be necessary to obtain a foreign patent. One option was to attract a locally based partner who could oversee the filing process. Alternatively, professional patent agents—who were in any case essential consultants for the serious inventor by this time—often handled international business directly. The specifications approved by the Patent Office in London's Chancery Lane typically included a number that appeared under the names of London patent agents rather than of inventors and were identified as "communications from abroad." Whatever the route to application, expeditious action was crucial because any form of prior publication in the destination country could compromise an eventual grant.[3] Receipt of the grant was, however, only the first phase of any act of intellectual property transfer. Each time a patent crossed national borders, the work of defining, defending, and exploiting the inventor's legal rights began anew.

Alexander Graham Bell gave early attention to his overseas rights. Because these were not covered by his agreement with the Bell Patent Association, shares in prospective foreign patents could be realized as a source of advance income while Bell was still a relatively penurious teacher of the deaf. In Canada during the winter of 1875–1876, he offered a half share in the foreign rights to his multiplex telegraph and telephone inventions to George Brown, a Liberal politician and proprietor of the *Toronto Globe*. Brown and his brother agreed to assume the cost of taking out patents and to advance fifty dollars per month until they had done so. In return, the Browns, worried that submission to the U.S. Patent Office might create prior publication under British law, insisted that Bell should not submit further American applications until after George cabled from England to confirm a filing there.

In January 1876, shortly after Bell sent his specification to Gardiner Greene Hubbard in Washington, he gave another to the departing Brown. The remainder of the agreement never came into force. Brown made no move to seek patents while in England. His brother would later recall that this failure resulted from unfavorable advice that George had received from a British electrical engineer.[4] Meanwhile, on February 14, Hubbard suddenly submitted the American patent application in Washington without word from either Brown or Bell. Hubbard's willingness to annihilate the Bell-Brown compact adds weight to the suggestion that he had advance intelligence about Elisha Gray's imminent caveat.[5] If so, the success of American telephony in the years that followed vindicated Hubbard's decision to put the domestic rights first. For the moment, however, the withdrawal of the Browns left the assertion of Bell's legal rights abroad one step behind the international publicity surrounding his invention.

The first sightings of Bell's invention in Britain thus predated the issue of a patent by some months, a time lag that imperiled the attempt to seek protection. In September 1876, the eminent scientist Sir William Thomson attempted the first British demonstration. Thomson had acquired a telephone from Bell after the Philadelphia Centennial Exposition in July. On his return, he exhibited the device to the meeting of the British Association at Glasgow, but could not elicit speech from it. Part of the instrument had been screwed down for the Atlantic crossing and had then been bent in transit; both features were assumed at Glasgow to be deliberate aspects of the design. As a result, and to the relief of a contrite Thomson, the "Glasgow instrument" was later held not to have anticipated Bell's patent. Thomson's display was, however, not the only potentially dangerous publication. That same summer, the journal *English Mechanic* had published a description of the telephone that would later force the Bell interests to give up part of their claim. It was not until December that the authorized version of the technology transfer took place, when Bell's London solicitor William Morgan-Brown filed what would become patent number 4,765 of 1876.[6] Another six months passed before Bell made the first moves to capitalize on his rights, by selling a portion of his British patent to the New England cotton merchant William H. Reynolds for $5,000.

Potentially the most important channel for introducing the telephone to Britain as a working proposition already stood open, in the form of the Post Office telegraph administration. Britain's telegraph networks had been nationalized in 1869 and placed under the control of the Post Office with an expansive remit to develop service.[7] Government telegraph officials were acutely aware of American leadership in the improvement of transmission technology: Joseph Stearns's duplex technique had already been taken up by the Post Office network, with Edison's quadruplex system shortly to follow.[8] The rapid strides of American telegraphy prompted an official fact-finding tour in the spring of 1877 by William Preece, a senior Post Office engineer and a prominent electrician in his own right. It was during this assignment that Preece met with Alexander Graham Bell and returned with the first working telephones to be demonstrated in Britain.[9]

By September 1877, Bell's partner Colonel Reynolds had entered negotiations with the Post Office to supply instruments for a government service. The official response was cautious. Leading Post Office engineers agreed that, for the moment, the technology was not practically viable, a view that strengthened as the Bell instrument proved incompatible with the existing district telegraph network.[10] The immaturity of the telephone, however, led officials to conflicting conclusions. Preece believed that the Post Office should acquire telephone rights immediately, before the working product commanded a higher price. The telegraph administration's other deputy chief engineer, the parsimonious Edward Graves, was more circumspect, noting that by established policy the Post Office refused to subsidize inventions in the development stage.[11] Even more than most private companies in the market for inventions, the Post Office was accustomed to contracting for finished products rather than sponsoring innovation. The department went so far as to refuse the use of its wires for inventors to test their own devices.[12] Post Office reticence was at least in part a question of principle: assuming the risk of an incomplete technology, in Graves's view, offended a government department's responsibility neither to speculate with public money nor to show favoritism toward any inventor.[13] The budgetary controls under which the Post Office worked eventually carried Graves's point for him. Although Chief Engineer R. S. Culley negotiated an agreement with Reynolds to lease telephones on preferential terms for a government private wire service, even this tentative approach foundered when the Treasury withheld permission for a contract.[14] The Post Office and Bell interests then parted ways while the government considered whether to assume legislative jurisdiction over telephony.

During the Post Office's deliberation, the primary concern of Bell and Reynolds was to reprise the publicity campaign that had preceded the launch of commercial operations in America. Bell, on his honeymoon, arrived in England in time to see Preece give the first demonstration of a working telephone in Britain, at the British Association for the Advancement of Science in Plymouth. The inventor himself lectured at the Society of Arts and the Society of Telegraph Engineers. For less technical consumption, Reynolds hired the American writer and performer Kate Field to pen articles and sing "Kathleen Mavourneen" over the wires, and Bell presented the telephone to Queen Victoria at Osborne House.[15]

These months of display had a serious part to play in the formation of a proprietary interest. Scientific opinion bore directly on the security of the patent rights, since reputation and a corps of sympathetic expert witnesses had to be laid by against the possibility of future litigation. Bell took full advantage of his genial association with Sir William Thomson, who represented Bell's entry point into the highest echelons of the British engineering establishment and provided Bell with letters of introduction to physicists and telegraph industrialists alike.[16] The intended audience for Bell's publicity undoubtedly included financial as well as scientific authorities, and as the Post Office negotiations lost momentum, the prospects of going forward with a private telephone venture waxed and waned with the public image of the patent. Unfortunately for Bell, a rash of hostile letters to the *Times*—some seizing on the possible prior publications that had occurred in Britain during 1876—seemed to weaken the potential value of his rights at the turn of 1878.[17]

Finally denied Post Office sponsorship and custom by the heavy hand of the Treasury, Bell and Reynolds turned to private capital. In March 1878 Reynolds sold 55 percent of the patent to a group of capitalists for £20,000, half of which would be paid after twelve months.[18] These investors in "The Telephone Company, Ltd. (Bell's Patents)" represented, Bell's secretary reported breathlessly, some seven million pounds of wealth between them.[19] Its leading lights included the merchant James Brand; W. Cuthbert Quilter, stockbroker and Member of Parliament, who soon became the single largest shareholder; the barrister and railway director John W. Batten; the cotton merchant George Dewhurst; and the financier Charles Morrison.[20] Morrison, one of the wealthiest Englishmen of his generation, may have been drawn into the syndicate through his association with the solicitor John Morris, one of the organizers of the telephone company and its first chief counsel.[21] Strikingly, no figures from Britain's powerful telegraph equipment industry figured among the telephone's backers. Gardiner Hubbard later rued the fact, lamenting that "it would have been difficult to have found the same number of able business men so little competent to carry on the business of the Company."[22]

Despite Alexander Graham Bell's presence in England during the first eight months of the Telephone Company's life and his assumption of a seat on the board, the patentee interest was not powerfully represented in the deliberations of the directors. Colonel Reynolds lacked executive ability and was quickly sidelined.[23] Bell made his most significant intervention in discussions of the business plan that the company should adopt, throwing his weight behind a version of the American model "according to which Telephones could be sold, under certain restrictions, for scientific and domestic purposes, while the main business of the Company would consist in the erection and maintenance of lines of Telephonic communication, remuneration being received in the shape of a fixed annual rental. The whole Telephonic system would remain the property of the Company, which would derive a permanent income from each line established."[24]

Bell's advice was an urgent criticism of the first general manager, one McClure, under whose leadership the company seemed to be drifting toward the mere sale of instruments. Bell did not find it "at all necessary that a Company should be organized, and a Board of Directors appointed, merely to superintend a shop for the sale of Telephones—to be managed by a man and boy!"[25] The other members of the board concurred in choosing the American business model and assented to the dismissal of McClure. Otherwise they were not generally impressed with Bell's directorial contribution, which waned over time. The inventor preferred to leave business questions to Adam Scott, a friend from his youth, who became acting secretary to the company in April. By October Bell's offhand relations with the board had deteriorated to the point where he resigned, citing "the gross mismanagement of the Company's business and the personal discourtesy with which I have been treated by the Board of Directors and the Acting Manager," and returned to the United States with his family.[26]

The patentee interest reasserted itself in 1879, however, in the person of Gardiner Hubbard. With the Boston capitalists ascendant in the new National Bell Telephone Company and his own credibility on financial matters in shreds among the American investors, Hubbard turned his attention across the water. At the time of Bell's marriage to Hubbard's daughter Mabel, the couple's patent interests in Britain, France, Belgium, Austria, and Germany had been placed in trust, with Gardiner Hubbard as the sole trustee. The chief promoter of early telephony in America thus became the key figure in the effort to project Bell's intellectual property rights overseas. Acting in the name of "Mabel's interest in the English, French and Austrian patents," Hubbard travelled to Europe determined to "put some life" into the apparently torpid English concern.[27] He found the Telephone Company struggling to make the transition from display to commercial operations: "Mr. Brand [the company's chairman] sends the Duke of Sutherland to see the telephone and all business is obstructed while the Duke and his friend see it. The next day comes the Duke's son and son-in-law and so all the time of the employees is wasted showing off the apparatus to Dukes, Marquises and Noblemen. They seem to be of more importance than business men."[28]

Picking his way through a divided and distracted board of directors, Hubbard embarked on two lines of reorganization. The first entailed creating district companies to develop service in provincial cities. Hubbard's first choice was Glasgow, but after consultations with the Telephone Company's capable Edinburgh agent and the major stockholder George Dewhurst, Hubbard's sights shifted to Manchester. With active support from Dewhurst, a Manchester resident, a pair of company agents set out to provide telephones for 100 subscribers on a budget of £1,500; if successful, they were to organize a local company capable of serving 500. Hubbard described the plan to Bell as "somewhat like our arrangements in America with changes to suit the circumstances."[29]

The trustee's second move was to reorganize the parent company, authorizing a major expansion of its capital and selling a tranche of 3,000 shares to raise money for immediate needs. The Bell trust's holding fell from about a fifth of the stock before reorganization to under 5 percent of the authorized capital, but Hubbard claimed he had more than doubled the value of the stake since the shares could now fetch around £2 10s each instead of £1. In keeping with Bell's wishes an attempt was made to place Adam Scott on the board as the trust's permanent representative, but some of the directors blocked the appointment and that role was taken instead by Richard Home, another Bell family friend.

Finally, Hubbard made sure that the company committed to a clear plan of action for developing the business. This plan mandated the division of London into fifteen districts under a metropolitan superintendent for canvassing and exchange-building purposes, and the appointment of at least three traveling agents to establish district companies in towns with over 100,000 inhabitants. A standard form of organization was proposed for the district companies: the issue of preference shares worth £2,000, and the parent firm taking half the ordinary stock.[30] When Hubbard left Britain for the United States after a busy four months (July–October 1879), he had laid the financial and strategic foundations for a more expansionist concern. In promoting a geographical extension of operations through part-owned local companies, he had also continued the importation—begun by Bell's insistence on renting rather than selling telephones—of the American company's evolving organizational model.

§

A few streets away from the Telephone Company's main office in the City of London, a transatlantic venture took place during 1879 that made Gardiner Hubbard's shake-up of the Bell concern look positively genteel. Crammed into the basement of a building on Queen Victoria Street, a posse of American engineers and workmen labored to make the Edison Telephone Company of London a working proposition. The prevailing historical image of this outfit was drawn by the playwright George Bernard Shaw, who briefly worked as an agent of the company. Shaw saw in his fellow employees a glimpse of the skilled proletariat of the United States. . . . They adored Mr. Edison as the greatest man of all time in every possible department of science, art and philosophy, and execrated Mr. Graham Bell, the inventor of the rival telephone, as his Satanic adversary; but each of them had (or pretended to have) on the brink of completion, an improvement on the telephone, usually a new transmitter. They were free-souled creatures, excellent company: sensitive, cheerful and profane; liars, braggarts, and hustlers; with an air of making slow old England hum which never left them even when, as often happened, they were wrestling with difficulties of their own making, or struggling in no-thoroughfares from which they had to be retrieved like strayed sheep by Englishmen without imagination enough to go wrong.[31]

Behind Shaw's romantic band of electricians lay another attempt to combine transplanted American intellectual property with English capital. In the United States, Thomas Edison had produced his carbon transmitter under the sponsorship of Western Union and had sold his portfolio of pending telephone patents to the telegraph company for $100,000.[32] In Britain, where Edison filed for a patent in July 1877, the inventor had a fresh opportunity to participate directly in its commercial exploitation.

By the beginning of 1879, Edison had compelling reasons to do so. First, the intervening months had established the carbon transmitter (or some imitation of it) as a vital component of high-quality telephones in the United States. Second, weaknesses in Bell's British patent made it significantly more vulnerable than the American original. Finally, Edison developed an alternative telephone receiver, his "electromotograph," which operated on a different principle from Bell's. The prospect of an instrument combining an electromotograph receiver with a carbon transmitter, taken along with doubts about the solidity of Bell's rights, appeared to offer a genuine chance of competition in Britain. Spurred on by this hope, the "Wizard of Menlo Park" and his associates set out to promote an independent Edison telephone in Britain by establishing a company in London and supplying it from the inventor's New Jersey headquarters.[33]

To pursue the exploitation of his British patent, Edison turned to Colonel George Gouraud, an American banker based in London who had earlier been involved in marketing Edison's automatic telegraph. The late 1870s saw Edison emerge as a figure of international standing on the strength of his phonograph, telephone, and electric light, and from that time onward Gouraud became a long-term collaborator in the transatlantic promotion of the inventor's rights.[34] Telephony was the first of Gouraud's major commissions, beginning with his appointment as Edison's agent in June 1878.[35] Other Edison associates soon arrived from the United States, bringing the necessary instruments and technical wherewithal to conduct experiments and prepare for public demonstrations.

The debut of Edison's telephone shows clearly the role of the leading scientific societies as the main interface between technologists and capitalists. Shortly after Professor William F. Barrett first introduced the carbon transmitter at the London Institution, Gouraud pressed Edison to send instruments for inclusion in a review of electrical progress by Professor John Tyndall at the Royal Society, advertising the opportunity as "the best possible opening for the financial campaign."[36] Thereafter the Royal Society became the frontline both for displaying the instrument and for soliciting capital. Tyndall's February 1879 lecture laid the groundwork for a further demonstration, this time at the house of the society's president and with the Prince of Wales in attendance. In negotiations on the fringes of this event, the Edison Company gained its first investor: Sir John Lubbock, a banker, parliamentarian, and naturalist who straddled the worlds of science, business, and politics.[37]

From Lubbock and the Royal Society, a web of social and professional connections spread out to the other founding shareholders in the Edison Telephone Company of London. Edward Pleydell Bouverie, who took the chairmanship, was another banker-politician and veteran of numerous Liberal governments as well as sitting with Lubbock on the Council of Foreign Bondholders. The architect Alfred Waterhouse was an associate of the Royal Society; his involvement in the Edison Company became a family affair with the appearance on the shareholders' list of J. Waterhouse, solicitor, and E. Waterhouse, of the accountants Price, Waterhouse & Co. Their family friend Sir Julian Goldsmid also took a stake, as did J. Lowell Price, head of Price, Waterhouse. The firm of Waterhouse and Winterbotham, solicitors, assumed the telephone company's legal representation. Other investors recruited by Gouraud included the former governor of Bombay, a partner in the mining firm Johnson Matthey, and the chairman of the London Stock Exchange.[38]

Unlike Alexander Graham Bell, Thomas Edison (via his agents) constructed an influential position within his British telephone enterprise.[39] The inventor held only one of the two hundred first-issue shares, but retained the right to name one director on a board numbering no more than seven. Edison's representative could veto the choice of general manager and call shareholder meetings at will. Immediate payment to the patentee for his rights was limited to a £5,000 advance; thereafter remuneration took the form of a 20 percent royalty on rentals or a sum equal to the dividend paid out by the company, whichever was higher. In addition, at the beginning of the year following the expiration of the patent Edison would receive half the then-value of the company's goodwill.[40] Gouraud and the capitalists reached agreement on these points relatively quickly, with the main elements in place by the end of May 1879. The bulk of negotiations leading up to the legal formation of the company in August concerned the right to exploit Edison's patent outside London.

The prospect of provincial development uncovered confusion about how much control Edison had ceded over his patent rights when he agreed to terms with the London investors. Colonel Gouraud returned to the city from an exploration of potential provincial markets to find that members of the still-pending London company had established local boards in Liverpool and Manchester, under the impression that they possessed a national franchise. This represented, as another Edison associate reported back to his employer in New Jersey, "a point of serious difference between you and them."[41] Gouraud found it necessary to compromise with the financiers by entering into a separate agreement, adding Lancashire (including Liverpool and Manchester) to the Edison Company's field of operations in return for a further £10,000 royalty advance to the patentee.

As he explained it to Edison, Gouraud conceded the formation of these local companies under the control of the London board for reasons of speed in countering Gardiner Hubbard's new Bell establishments.[42] Nevertheless, the colonel still thought Edison's interests best served by a staunch defense of his right to grant more franchises independent of

London, and promptly embarked on the promotion of a separate company in Glasgow.[43] The final agreement between Edison and the capitalists accordingly described a very restricted grant: the Edison Telephone Company of London had the right to operate within the East Central postal district of the city—without doubt the country's prime market—and to add other London areas to the franchise in return for £5,000 in advance royalties per postal district. With the exception of Lancashire, Edison reserved the right to establish new companies elsewhere in Britain.[44]

The launch of the London firm in August did not put an end to the district company question. Instead the issue simmered on late into the year, with Gouraud sustaining his efforts to keep regional development independent of the London board. As time went on, the tension between the patentee interest and the company was further complicated by an emerging struggle over who represented Edison in Britain. Gouraud held a power of attorney from the inventor, arranged at the time of the Lancashire negotiations and renewed when the colonel joined the board.[45] From July, however, Edison's Menlo Park lieutenant Edward H. Johnson took charge of technical affairs at the London company, and his authority with the other directors grew as he dictated preparations for the impending patent fight with Bell. Johnson sided with the rest of the board in seeking to coordinate Edison telephone promotions in Britain through the London company. Together, they eventually overrode Gouraud's objections to a settlement in which Edison and the company jointly exercised voting control of all new district operations.[46] Gouraud's insistence on stand-alone companies drew no credibility from his Glasgow project, which crumpled amid investor acrimony.[47] By the close of 1879, the Edison enterprise had staggered to a parent-subsidiary structure approximating that of its rival, but the full costs of internal strife and delay were increasingly clear as the Bell interests consolidated their lead in the field.

The Edison Telephone Company of London had grown out of a conscious attempt to avoid repeating the patent deadlock created in the United States. Thomas Edison had gone so far as to develop an "electromotograph" receiver that used electrochemical charge instead of Bell's electromagnetism, in order to free his patented carbon transmitter from bondage to a Bell-type receiver.[48] Nevertheless, a long staring contest took place in the first months of competition while the rival companies took legal advice and calculated whether their respective patents would hold. Each new piece of favorable or unfavorable opinion provoked a dramatic collective mood swing among directors, executives, and technical staff. Hubbard reported an episode of directorial gloom in August 1879 when the Telephone Company's solicitors decided that Bell's patent could not be maintained. At the Edison Company, personal loyalty to the inventor kept the American employees generally convinced of their patent position, but even the confidence of the fervent Edward Johnson could be shaken.[49] As in America, it was impossible for the protagonists to predict whether financial and commercial factors would decide the legal situation or vice versa.

Telephone competition barely left the starting gate in Britain, in contrast to the continental-scale race under way on the other side of the Atlantic. Whereas American Speaking Telephone had drawn on the massive technical and corporate resources of Western Union in pursuing the Bell Company, the Edison concern lacked the means to catch up with its opponent's head start. Delays beset Edison's London operation from the beginning: a wait for the electromotograph, technical problems with the instruments, a long hiatus while unreliable sections of the carbon transmitter patent were disclaimed in Patent Office proceedings. Between the time lag in establishing the technology and the sluggish commercial start by both major companies, British telephone development was over a year behind that of the United States already. The firms that faced each other in 1879–1880 were miniature by comparison to the American concerns.

As a result, no serious competitive battles took place in the cities. An advance order by Gouraud for two thousand instruments proved highly optimistic; by the end of 1879, the three Edison exchanges in London had fewer than 170 subscribers against the Bell interests' 200. In the provinces, Telephone Company exchanges had begun to operate in Liverpool, Manchester, Edinburgh, Birmingham, and Bristol, albeit with no more than a handful of subscribers each.[50] It is not clear whether the Edison enterprises canvassing in Liverpool, Manchester, and Glasgow had yet succeeded in constructing facilities. In the absence of significant confrontation in the field, little can be deduced about the effect of Edison's

£12 annual rate as against the Telephone Company's £20 charge. Contemporary comparisons focused on quality, where Bell had a slight lead. As in America, the Bell side brought in Francis Blake's transmitter early in 1880 to cancel out Edison's superiority on that front. Meanwhile the electromotograph receiver proved a major drag on Edison's instrument.[51] This component had to be continuously cranked with one hand and thus prevented the user from writing down messages. The whole Edison device was also notoriously loud. George Bernard Shaw described it as "a telephone of such stentorian efficiency that it bellowed your most private communications all over the house instead of whispering them with some sort of discretion."[52]

The future of this infant industry was clouded at the end of 1879 by the actions of the Post Office. Lawsuits prepared against both the Bell and Edison companies asserted Post Office jurisdiction over telephony under the parliamentary act that had nationalized the telegraphs. The government chose Edison for its test case, and the company had to fight on an unexpected legal front to demonstrate that the telephone was not, in fact, a telegraph.[53] Still more dangerously, the Post Office began to play the two telephone firms off against each other. Arnold White, secretary of the Edison Company, uncovered advanced negotiations between the Telephone Company and the Post Office under which the Bell interests would have bowed to regulation and a 10 percent levy on receipts in return for government preference and patronage. At the last moment White persuaded John Batten, "the man on the Bell board with influence and brains," to make common cause against this divide-and-rule strategy.[54] It was a logical step, as Edward Johnson noted, from this tactical cooperation to a merger of the companies.[55] The Post Office threat in any case encouraged both boards to contemplate strategies of market stabilization and investor reassurance, in order to shore up their stock value and position themselves for the possibility of state purchase.

Thoughts of amalgamation began to steal into Edison Company thinking in November, when Edward Johnson remarked that preserving rather than attacking the Bell patent might repay dividends in the event of a merger. The Telephone Company then began to dangle the prospect openly, with a corrosive effect on Edison investors' fighting spirit. "I cannot alone support all these weak knees," Johnson complained.[56] Two approaches were made directly to Thomas Edison in New Jersey. The first came from the Western Electric manager DeLancy H. Louderback, Western Union's former agent in the southern United States and one of the two principal designers of the competition-ending agreement in that region. On unspecified authority, Louderback proposed a consolidated British company in which each firm would receive stock equal to the cost of its plant.[57] Less than a week later a second approach was made by an English representative of the Telephone Company.[58]

Both met with a summary rebuff from the inventor. Thomas Edison continued to believe that Western Union had sold his telephone patent short in America, and would not allow his rights to yield again before what he considered Bell's weak British claims. "Their patent aint worth a cent," he scrawled on Louderback's proposal; "the one of Blake is a d .. d piracy and will be stopd. When they notified me to stop using Bell's receiver in England I did. Now I propose to make them stop using my carbon tel if I lose every cent and never make a cent in Engd that is a subject I am sure on."[59] Clearly, Edison's personal involvement in the London company made conciliation with the Bell interests harder than it had been in the United States. The inventor's stake in the final vindication of his patents was of an entirely different order to the purely financial investment of the "weak-kneed" shareholders.

Only two months later these proposals looked like missed opportunities for the Edison interests. A fresh share issue by the Telephone Company early in 1880 cruelly exposed the Edison concern's twin shortages: money and time. With an enormous influx of capital gained from placing large amounts of stock at a 100 percent premium, the Bell leadership professed "to care nothing for patents now."[60] The Edison Company's legal advisors were unanimous that such a financial advantage would allow the Bell side to prolong any litigation while sweeping their opponents from the field; Edison would never get his day in court. At the same time the merger offer was withdrawn.[61] In the face of this crisis, a split emerged at the Edison Company. Gouraud pushed for a public stock offering, whereas Bouverie, White, and Johnson disdained the suggestion, protesting that a "stock jobbing movement" would turn the company into a speculative vehicle and see Edison's reputation "sacrificed on the altar of Mammon."[62] Their solution was to amalgamate the Edison companies in London and Glasgow and raise more capital from existing investors.

In two ways the fate of the Edison enterprise now hinged on the operation of the patentee interest within the company. First was the need to seek a renegotiation of the inventor's royalty. Only after hard bargaining did Thomas Edison exchange his 20 percent share of gross receipts for £10,000 and a profit-sharing arrangement.[63] Second, a bitter struggle for authority between Edison's agent and the other principal officers deadlocked decision making at board level, with Gouraud obstructing Bouverie and Johnson for the remainder of the Edison Company's life. Relations between the board in London and Edison's official representative finally deteriorated to the point where Johnson lobbied Edison to revoke Gouraud's power of attorney, the board rejected the colonel's authority, and Chairman Bouverie opened a channel of communications direct to Menlo Park.[64] If the internal refinancing of the Edison enterprise had possessed any hope of success, it disappeared in these fights—victim of a governance structure in which the inventor's distant authority, exercised only sporadically by Edison himself, allowed his agent to thwart the board.

Amalgamation between the Edison and Bell companies was negotiated in spring 1880, motivated on the Bell side most likely by the desire to achieve monopoly status before any Post Office purchase, and on the Edison side by resignation to financial defeat. The Edison enterprise took the junior position in the consolidated company, yet shares were allocated to reflect the value of plant and finances alone; the two sides' patents were given parity.[65] The new firm, which labeled itself "The United Telephone Company Ltd. (Bell's and Edison's patents)," began with an authorized capital of £500,000, of which £200,000 in fully paid stock went to shareholders of the Telephone Company, £115,000 to Edison owners, and £85,000 was offered for sale to existing investors. James Brand of the Bell interests took the chair, with Bouverie as vice chairman. James Staats Forbes, the railway executive who had brokered the merger, now joined the board and became a powerful force in the company. Meanwhile the role of the inventors and their agents faded. Edison's representatives dissolved the inventor's personal stake, though Gouraud kept his hand in telephone matters at a succession of other firms. Hubbard sold off most of the Bell trust's shares in the United Telephone Company within a year.[66] After the fractious period of transatlantic interventions, investors looked forward to greater harmony with the capitalists in charge.

§

Viewed in a comparative context, the telephone patents demonstrate the continuous making and remaking of the patent interest in a technology. From the earliest stage, the selection of one proprietary course over another—such as the telephone remaining separate from national telegraph operators, rather than being absorbed by them—determined the commercial form taken by the technology. But even later, whenever patents became a factor in the internal governance of firms or an instrument for opening business in new countries and new territories, the issue of ownership pushed to the fore.

Within countries, the stance of incumbents toward new technologies had a powerful early say over the commercial context of development. Decisions taken by Western Union and the Post Office effectively forced telephone patentees in the United States and Britain to pioneer a new and separate industry, although the absorption of voice transmission into existing communications networks as a local complement to longer-range telegraph service was technically and economically quite conceivable.[67] Incumbents had a range of strategies available in the face of a new patented invention: they could adopt the device for their own use, or reject it into the hands of others, as the telegraph concerns at first dismissed the telephone. Alternatively, since the strength of the rights involved could not be taken for granted, incumbents had the option of trying to undermine the patent.

The decision to pay for, to ignore, or to attack a patent was a judgment on the intellectual property as well as on the technology as such. Much has been written on the hubris that caused the telegraph monopolies to regard Bell's instrument as a "scientific toy," but these were institutions that took decisions about patent acquisitions with a hardheaded view of what their money bought.[68] Not only Bell's boxy prototypes but also his legal rights were unproven in 1876–1877. In light of this uncertainty, Western Union's decision to decline Bell's invention seems less dismissive of the technology; the company, after all, continued to sponsor Edison's research. Orton and his colleagues knew full well that there would be many later opportunities to invent around, to acquire, or to destroy any patent in their path. Conversely, officials at the British Post Office were more concerned with the possibility that Bell's rights would gain value over time, making government purchase more difficult, and saw the perceived weakness of the patent as a window of opportunity to act. William Preece advocated a swift arrangement with Colonel Reynolds for just that reason, before the Treasury summarily ended discussions. To be sure, complacent assumptions about speaking telephones themselves, such as the expectation that business users would continue to prefer a written record, did help to sway the telegraph executives. But any retrospective assessment of these men's failure to seize on Bell's invention should concede the uncertainties of the patent situation.

If the patents helped to decide who commercialized the telephone, they had an equally significant effect on how the telephone business took shape. Relationships between capitalists and patentees took center stage in the expansion of the embryonic patent-based concerns. Late nineteenth-century inventors were often active entrepreneurs, holding a formal position as officers or principals of the firms that developed their innovations.[69] In such cases, renegotiating the patent-holder's status as the enterprise expanded could be a tense process. Thus the first entry of Boston financiers into the Bell telephone enterprise became a point of intellectual property transfer, with Gardiner Hubbard effectively relicensing New England in order to realize fresh capital and at the same time to reinforce the patent owner's prerogative by ring-fencing the newcomers within a specified territory. The same issues were at stake in George Gouraud's battle with the Edison investors in Britain: the patentee's ability to subdivide legal title allowed the piecemeal exploitation of Edison's rights, and hence enabled the inventor and his agent to assert independence. These examples show the potential for a patentee interest to force compromises in the strategic direction of the firm as a whole. On the other hand, the role of patent-owning entrepreneur could also supply a drive otherwise lacking, as Gardiner Hubbard demonstrated when he initiated the provincial expansion of the Telephone Company. In the final analysis, whether the results were positive or negative, each of the young telephone companies experienced the transfers and assignments of intellectual property as an ongoing process that altered the internal balance of power in the firm.

International transfers threw open not only the composition of the proprietary interest in the destination country, but the survival and scope of the patent itself. Time, distance, and the variability of national legal systems all conspired against the foreign (especially transatlantic) patentee of the 1870s, especially when descriptions of the invention had already begun to circulate. Patentees with international aspirations faced a complex task in securing protection and local commercial partners. As they did so, however, the reproduction of intellectual property from one country to the next formed a kind of legal shadow to the transmission of technical knowledge and material artifacts—"international technology transfer" as it is traditionally understood.[70] The case of the telephone suggests that the movement of property rights could be every bit as important in the diffusion of a technology as the movement of machines or ideas.[71]

Some further comparisons between national telephone industries serve to illustrate the point. On the one hand, Britain, France, and Canada all granted patents equivalent to those of Bell and Edison in the United States. In each of these countries, early private operators combined or pooled the basic grants to create a parent telephone company of national scope. Circumstances were by no means identical—France nationalized the telephone company in 1889, whereas Bell's (though not Edison's) rights were struck down in Canada in 1885—but the organization of telephone enterprise was strikingly similar during the first few years of the technology.[72] In Germany, by contrast, the attempt to manage intellectual property transfer failed, largely because of unfortunate timing. The year 1877 marked the enactment of Germany's first Reich patent law, following years of lobbying led by Werner von Siemens. The law went into force in July 1877; Bell's patent application in August came too late to preempt the circulation of telephone designs in the technical press. As Siemens informed Bell in November, telephone manufacture in Germany had no need for his permission.[73] The absence of a controlling patent in private hands cleared the way for Germany to establish a state postal monopoly supplied by competing manufacturers.[74] A third model of telephone development prevailed in Sweden, which lacked a patent law in the 1870s. Despite the establishment of a Bell franchise, the telegraph engineer L. M. Ericsson was able to develop his own version of the basic device and institute local competition in Stockholm. What must have seemed a small gap in Bell's patent coverage at the time eventually became a significant opening in the history of telephone technology: Ericsson's firm became a leader in innovation and export, and Stockholm by the early 1890s had the highest number of telephones per capita of any city in the world.[75]

Elsewhere, the relationship between patents and early telephone service was less straightforward. In the smaller countries of Europe (including Switzerland and the Netherlands, which also lacked patent laws), telephone provision was structured predominantly by local government licensing and by municipal cooperatives.[76] In markets such as these, the proprietary aspect of the industry tended to take the form of multinational companies operating local concessions, either alone or as part of a joint venture. A number of vehicles appeared after 1880 to promote "Bell" or "Edison" telephony either with or without legal protection in the countries they covered. Gardiner Hubbard organized the International Bell Telephone Company in New York, which took over a number of national Bell patents in Europe, and shortly afterward Hubbard and the American manufacturer Western Electric jointly established the Bell Telephone Manufacturing Company to supply the continent from a centralized manufacturing facility at Antwerp.[77] Meanwhile Edison and his associates formed an Edison Telephone Company of Europe, which did most of its business in Belgium and Portugal. Outside Europe, the Continental Telephone Company, headed by the American Bell director George Bradley, was chartered in Massachusetts and began to establish systems under Bell license in South America. The broadest horizons of all belonged to the Oriental Telephone Company, again promoted by Hubbard in concert with a group of English capitalists including the "Cable King" of submarine telegraphy, John Pender. This concern undertook to develop Britain's colonies, the Middle East, the Far East, and Australasia.[78]

Seen amid the variety of fates that befell telephone patents across Europe and the world, the proprietary enterprises developing in the United States and Britain appear far more similar than they were different. Britain not only received the transfer of intellectual property from America more quickly than elsewhere, but also produced the closest foreign simulacrum of the original Bell Company's business model—and then a second one, in the form of Edison's venture. Of course, this outcome was no coincidence; most of the early transatlantic parallels followed from the personal role of America's pioneering telephone inventors and entrepreneurs in launching the technology in Britain under their own patents. The content of the patents themselves was another matter. One might assume that the same technology, promoted by the same inventors, would generate the same claims of ownership on both sides of the ocean. What came to pass was quite different.

CHAPTER 6

Patent the Earth

IN 1913 THE ROYAL SOCIETY AWARDED its Hughes Medal to Alexander Graham Bell, "on the ground of his share in the invention of the Telephone and more especially the construction of the Telephone Receiver."[1] According to a contemporary observer, this relegation of Bell from sole inventor to ensemble player reflected a generally lower appreciation of his role in Britain than in the United States.[2] Such national disregard may seem surprising. It was natural for Germany to claim the original invention for Philipp Reis and for Italians to hail their own Innocenzo Manzetti, but the Scots-born Bell remained Britain's one serious claimant to the honor.[3] In fact, the Royal Society's allotment of historical credit followed the legal situation. Bell's British patent controlled only the telephone receiver effectively. As a consequence, Thomas Edison's patent not only became the key to control of the transmitter, but increasingly featured as the basis of the telephone monopoly. Had Bell's patent controlled the telephone as completely as it did in the United States, he would surely not have had to share the laurels—yet share them he did.

The whole legal theory of the telephone invention in Britain differed from the understanding that prevailed in the United States. Bell's "undulatory current" did not become the dominant concept in assigning ownership of the invention, causing Britain's leading electrical journal to scoff that "America is . . . the only country where this preposterous proposition [Bell's fifth claim], as a broad claim, irrespective of any form of apparatus, is allowed to stand."[4] Identical technology thus translated into different configurations of patent rights in the two countries. The difference in the way that national patents ended up being drafted and interpreted is another reminder of the contingency of patent rights—a sign that the mapping of legal claims onto a particular invention could happen in a myriad of ways. It also provides a counterfactual scenario. The British story is a window onto what might have happened in the United States if Bell's claims had been more narrowly understood and the technical distinctions between types of telephone had actually mattered.

Comparing the two countries' experiences highlights a variety of structural differences and similarities. Despite a nominally shared heritage, the two nations' patent and legal systems diverged in many respects. Yet two factors worked in Britain, as in America, to determine the outcome of the telephone patent fight. One was successful lawyering: the choices and strategies pursued by the United Telephone Company's lawyers proved capable of constructing patent dominance out of initially unpromising materials. The other was judicial support for the idea of the heroic inventor, which became one of the animating justifications for the patent system and came to pervade its doctrines. Thanks to these two factors, the two national stories reached similar conclusions. By the end of a string of court cases in the 1880s, the United Telephone Company had successfully closed out all its competitors and achieved a monopoly of basic telephone technology comparable to that of American Bell. The striking thing is that, although the respective end-points had so much in common—a pioneer inventor, a broad patent claim, a monopoly firm—in Britain, the inventor and the claim in question were entirely different.

§

The patent systems of Britain and the United States differed in scale, substance, and significance. America's patent regime by the later nineteenth century was democratic, bureaucratic, powerfully entrenched, and prominent in the business of the federal courts. Britain's patent system was altogether less imposing—weaker in administration, thinner in judicial support, and for a time threatened with outright abolition. The differences had long roots. From its beginnings in 1790, the American patent system had commanded widespread political and judicial legitimacy as a tool of industrial encouragement. Its British counterpart had evolved more haltingly as a regime of royal privileges and, as a result, had a more dubious legal and political status. Until the 1830s, residual suspicion of monopolies led English courts to appear generally skeptical toward patents. Many judges were known to treat patents harshly, "as if the object was to defeat and not sustain them," and to invalidate inventors' rights based on trivial failures to comply with the legal requirements of the grant.[5]

Administratively, progress in Britain was slow. Throughout the first half of the nineteenth century, the act of obtaining a patent remained an expensive and byzantine task involving at least seven different government offices. The process was memorably mocked by Charles Dickens as a forced march from one fee-hungry royal official to the next, starting with the lord chancellor and the home secretary and ending with "the Deputy Clerk of the Hanaper, the Deputy Sealer, and the Deputy Chaffwax."[6] By midcentury, the embarrassing state of the law had become too acute to ignore. The Great Exhibition of 1851 crystallized the efforts of reformers, who could point to the lack of effective legal protection for the inventions and manufactures displayed amid such pomp at the Crystal Palace.[7] The following year brought the first major reform of patent legislation since the 1624 Statute of Monopolies. The 1852 Patent Act reduced the initial filing cost for nationwide patent rights from over £300 (or around £100 for England and Wales alone) to £25.[8] A new Patent Office was created to handle applications and maintain an archive of patents, though its role was pure registration: the office did not examine applications for either novelty or validity. Responsibility for the whole apparatus lay with a group of senior judges and legal officeholders acting part-time as the commissioners of patents.[9]

Ironically, the same reforms that created something more like a modern patent system also spurred a movement in opposition to the patent law itself. The beleaguered reign of the commissioners, lasting from 1852 to 1883, witnessed repeated commissions of inquiry into the workings of the system and a lengthy campaign for the outright abolition of the patent law.[10] Abolitionism was a mixture of different economic and political impulses. One was a set of practical objections and grievances. The reform of 1852 had failed to end complaints about the cost and complication of obtaining the grant. At the same time, reduced fees created a large enough rise in the number of patents for some industrial users to find the growth of protected inventions obstructive. Prominent critics from industry, including the famed engineer Isambard Kingdom Brunel and the armaments magnate William Armstrong, stepped forward to assert that the patent system "impedes everything it means to encourage, and ruins the class it professes to protect."[11] A second route to abolitionism ran through imperial political economy. The 1852 Act specifically exempted the colonies from patent protection, apparently an attempt to encourage the mechanization of West Indian sugar refining, but a decision which enraged domestic sugar refiners and created one of the more influential lobbies calling for abolition.[12] The third impetus was free trade, the ruling economic ideology of nineteenth-century British liberalism, which seemed to many commentators to be incompatible with a system of monopoly patent grants. This viewpoint gained purchase in other European countries at around the same time. Free-trade principles were influential in delaying the adoption of a common patent system for the German states and in leading the Netherlands to abolish its patent law in 1869.[13]

At its high-water mark in the late 1860s and early 1870s, the patent abolition movement had the support of industrialists and men of science, of the *Economist* and the *Times,* and of high-ranking politicians including the attorney general and key figures in the House of Lords.[14] Yet abolitionism spurred a counteroffensive, not only on behalf of the patent system, but on behalf of the idea of invention itself. Inventors, scientific societies, patent professionals, and their allies mobilized to promote the idea of heroic invention and to link it to the rewards granted by the patent law. One key propagandist was Bennet Woodcroft, an inventor and patent agent who in 1852 became clerk to the commissioners of patents, effectively the chief of the new Patent Office. The energetic Woodcroft pursued favorable publicity for inventors at every available turn, publishing pamphlets, opening a Patent Office Museum, writing biographies of inventive pioneers, and compiling the historical records of the patent system going back to 1617. Woodcroft was in close touch with the writers who made the striving inventor a central trope of Victorian popular literature—above all, Samuel Smiles, whose 1859 paean *Self-Help* was one of the great best sellers of the day.[15]

The defenders of heroic invention were able to tap into strains of cultural celebration that had already started to elevate technology within the firmament of British society. This project was relatively recent. Before the mid-nineteenth century, British invention lacked a pantheon; inventors were obscure characters, and many of the figures later seen as pioneers of the Industrial Revolution languished unremembered in forgotten graves. That began to change in the 1830s, beginning with James Watt, who posthumously became a one-man symbol of the age of steam. Watt's apotheosis reflected nothing less than a gradual shift in portrayals of national identity, one which elevated technological achievement and the rising industrial classes to stand alongside the military-aristocratic heroes of yesteryear.[16]

The greatest outpouring of inventive celebration, though, came in the third quarter of the nineteenth century. Biographies and statues of past inventors appeared with unprecedented frequency. Alongside Watt, other heroes of the Industrial Revolution were commemorated in their provincial resting places and in Westminster Abbey. This period saw the peak of the debates over the patent system, and the defenders of the patent law both drew on and contributed to the heroic corpus.[17] To be sure, they did not go unchallenged. The abolitionists put forward an alternative vision of invention as a cumulative, collective process, characterized by simultaneous and incremental improvement rather than by acts of individual genius. Such a view was at a natural disadvantage, though, against the proponents of the heroic narrative and the cultural capital that they could muster.[18]

In the end, the abolition movement fell short. The election of a Conservative government in 1874 removed some of the most active Liberal abolitionists from office, and criticism of the patent law was channeled instead into reform.[19] Statutory reform, when it came, created a more regularized and accessible system. Under the Patent Act of 1883, initial filing fees fell to £4, flinging open the doors of the patent system to a greatly increased number of applicants. Patenting approximately doubled under the lower fee schedule. The new statute finally instituted professional examination for validity, though unlike their American counterparts, British patent examiners did not check the novelty of patent applications until early in the twentieth century. This minimal quality control did little to offset the large quantity of potentially invalid or evidently low-value patents. Both before and after the 1883 Act, some two-thirds of patents issued were abandoned after three or four years when the first upkeep payment fell due.[20] An official inquiry in the early 1900s revealed that more than 40 percent of patents issued were anticipated by earlier grants.[21] Mere receipt of a patent under this system conveyed little certainty about the validity of the rights it described. That problem was left to the courts.

Patentees who entered the courts found that the world of patent practice was a small one. Its limited scale reflected in part the limitations of the judicial system generally. Civil lawsuits concentrated in London, with businessmen outside the capital left to rely on an inefficient system of assize hearings by judges on circuit. Late nineteenth-century litigation at least represented an improvement on the procedural complexity and delay of earlier decades—particularly the legendary stasis of the Chancery courts—but still suffered from deficits of competence and capacity. A frustrated business community increasingly turned to arbitration, a trend that was arrested only by the creation of an ad hoc "commercial court" of appropriately qualified judges in 1895.[22]

In this environment, few patent cases seem to have taken place. From his own records and a survey of his colleagues, the lawyer Theo Aston calculated that some 109 cases had commenced in the English courts between 1865 and 1870 (an average of just over 18 per year) and that 65 (around 11 per year) had reached judgment during the same period.[23] As Aston pointed out, these numbers were tiny when compared to more than thirty thousand suits filed in the superior common-law and equity courts during these years.[24] Another authority put the incidence of patent litigation even lower in the mid-1870s, at around nine cases per year.[25]

Some increase in litigation followed the reforms of 1883 and the subsequent growth in patenting, though how much is unclear. The record of published judicial opinions does reveal a few clusters of litigation after the 1883 Patent Act: 8 telephone cases (all brought by the United Telephone Company) out of 57 total reports in 1886; a concentration of phonograph cases in 1894 (8 of them brought by the Edison Bell Phonograph Company); another group relating to the Welsbach gas lamp in 1900–1901; and 2 bursts of pneumatic tire litigation in 1896–1897 and 1901–1902.[26]

The low volume of patent business in the courts was nevertheless enough to attract an identifiable patent bar. A few technically literate barristers developed patent practices in midcentury, notably William Hindmarch, author of a leading textbook; William Robert Grove, a noted physicist best known today as the inventor of the fuel cell; and Thomas Webster, who led the field from the 1840s to the 1870s.[27] The patent bar of the next generation—literally the next, in the case of Webster's son Richard—was still a small group. Few of its members in the 1880s specialized completely. Theo Aston, a pupil of Thomas Webster's, was an exception: he took patent cases exclusively, accepting a marginal position among commercial barristers in return for prominence in his area.[28] Other lawyers took a broader mix of cases, most often combining patent work with Chancery practice or commercial law.

Among these men, the leading figures drew their standing from a variety of sources. A few were mathematically or scientifically trained, including John Fletcher Moulton, a fellow of the Royal Society on the strength of his own electrical research, and T. M. Goodeve, who held professorships in engineering and mathematics in addition to his legal practice.[29] Of the remaining barristers who did substantial business, perhaps a half dozen were sufficiently specialized to author volumes of reports or treatises on the patent law. Finally, the attorney general and solicitor general were typically valuable representatives in patent cases. No occupant of these offices proved more important to the patent arena than Richard Webster. A brilliant advocate who attained the rank of queen's counsel at a young age, Webster followed his father's footsteps to a sizeable patent practice.[30] While attorney general under three governments between 1885 and 1900, Webster continued his private practice and led many of the largest cases in the courts. The *Electrical Review* rated him "an awe-inspiring blunderbuss" in litigation.[31] In the 1890s, during Webster's second spell as attorney general, the government requested that he set aside private practice to focus on official duties, whereupon leadership of the patent bar passed to Moulton.[32] Almost every major patent case of the late nineteenth century involved one or both of these two men.

Alongside these barristers was another type of patent specialist in the courtroom: expert witnesses. The ability to explain technical details to the court and to lay out the prior art in support of one party's arguments was so highly prized that, while a scientist from the relevant field might be called in a case, it was common to employ a generalist consulting engineer whose true specialty lay in expert courtroom testimony. The back-and-forth between these men and the luminaries of the patent bar turned patent litigation into "a series of duels between counsel and the opposing witnesses in which the opponents are on very equal terms, since the counsel are generally well acquainted with the scientific side of the matter, while the witnesses are perfectly familiar with the rules of procedure and the principles of law involved. Further, each party is from long experience well acquainted with the other's mentality."[33]

Such personal familiarity arose because, like the patent bar, the corps of expert witnesses had an inner circle that was near-ubiquitous in important cases. John Imray, a patent agent, was a regular on the stand during the later nineteenth century, as were Dr. John Hopkinson, a consulting engineer and professor of electrical engineering, and Sir Frederick Bramwell.[34] When on one occasion Imray, Hopkinson, and Bramwell all appeared in the same case, Mr. Justice Denman was moved to note the "conflict of testimony between such high authorities."[35] Of the three, Bramwell was the most highly regarded by judges and counsel. Usually describing himself as a civil engineer, Bramwell occupied the heart of the British legal-scientific-engineering establishment, serving during the 1870s and 1880s as president of the Institute of Mechanical Engineers, the Institute of Civil Engineers, and the British Association for the Advancement of Science, and as chairman of the council of the

Society of Arts. He also happened to be the brother of a distinguished judge.[36] Bramwell's authoritative depositions in patent cases included appearances for Bell and Edison in the telephone litigation and for Edison in the electric lamp cases, making him one of the most important interpreters of electrical technology in the British courts.

Amid this collective accretion of patent expertise by lawyers and other courtroom professionals, the group whose contribution is perhaps hardest to evaluate is the judiciary. The slim body of reported case law, especially before 1884, ensured that few British judges could build up a profile as leaders in this area. Certain structurally important offices gave their occupants greater exposure to patents. Among the six commissioners of patents were the lord chancellor and master of the rolls (the two top Chancery judges) and the solicitor general and attorney general (both positions that were common stepping-stones to the bench). But few judges before 1900 were closely identified with the field.[37] One who combined an ex officio role with a special personal interest was Sir George Jessel, master of the rolls from 1873 to 1883. Jessel was "working head of the Patent Office" in his capacity as a commissioner under the 1852 Act, and at the same time was a moving force in patent jurisprudence, in large part through his authority in the construction of written documents.[38] Perhaps the only judge who shared Jessel's stature in patent questions during the 1880s was W. R. Grove, the scientist and former patent barrister who sat as a judge of the Court of Common Pleas from 1871 to 1880 and of the Queen's Bench from 1880 to 1887.[39] Notwithstanding these occasional experts, though, the leading rulings in the patent law issued from a disparate group of judges and did not—unlike in America—loom large in the business of any court.

Across the period of the major telephone patent cases, then (roughly 1881 to 1886 in Britain), the leading repeat players tended to be on the litigants' side of the bench, rather than the judges'. Legal advice and support for patentees may not have been as widely available as they were in the United States, but that mattered little to such serious and well-resourced parties as Bell or Edison, who could and did recruit the best talent. In fact, the limited size of the patent bar and of the cadre of leading experts added a further element of advantage for powerful litigants because it was possible to monopolize the advice and arguments of the most authoritative figures. This imbalance of power would prove crucial, and perhaps decisive, in the resolution of the telephone cases.

§

Comparison of the telephone patents in Britain and America begins with the written documents themselves. Both Bell and Edison modified their original specifications in seeking British rights: a move that, given the sensitivity of patent interpretation to the precise language of the document, necessarily altered the way in which courts would understand the two inventors' rights. Even more important were the amendments made to the British patents after these were granted. The content of the basic patents remained in flux well into the lifetimes of the first telephone companies, resulting in much-altered claims that set the proprietary status of British telephony on a different course.

Some of the greatest complications for Bell's legal situation stemmed from the simple matter of time lag. Bell had prepared one British patent application before his American filing: the specification that George Brown took to London in January 1876 but never acted on. By the time he prepared a second version in October of that year and had it submitted in London by the patent agent William Morgan-Brown on December 9, fully ten months had passed since the American application.[40] Bell's new document took full account of developments in the interim. Speech transmission had never actually been achieved at the time of the American application (which mentioned only "vocal or other sounds") but was a working reality when Bell drafted the British specification. The later patent was thus able to describe a practical telephone based on Bell's "magneto" design. By sweeping in all his inventions to date, Bell produced an omnibus British patent. It consisted of six "plans," the first three of which described systems of multiplex telegraphy, the fourth and fifth the means of speech transmission (with and without battery power in the circuit), and the sixth a form of "autograph" telegraph for transmitting writing. No fewer than eighteen formal claims followed. Four of the eighteen came word-for-word from Bell's original U.S. telephone patent; seven were shared with his second major American grant, which described the magneto telephone; and two appeared in lesser patents.

The one striking exception to the direct replication of language was the fifth claim of his original telephone patent—the single sentence that became the foundation of patent monopoly in the United States. In Bell's British grant, the fifth American claim was broken up into two different, wordier claims that distinguished between electric undulations corresponding to "variations of density produced in the air by the said sounds" (claim seven) and those corresponding to "the velocity and direction of motion of the particles of air during the production of the sounds" (claim eight).[41] It is not clear what Bell thought he would gain by this distinction. More than anything, the change reminds us that Bell did not initially place particular emphasis on his fifth claim; it was Chauncey Smith's litigation strategy in the Boston circuit court that had privileged that part of the patent.

Drafting choices aside, there were more fundamental reasons why Bell's British rights could not rest on a single catchall claim. The delay in filing a British application had created opportunities for prior publication that sharply changed the prospects of the patent. Even as he prepared to mail his final specification to London in October 1876, the inventor fretted that "I shall wish it good luck although I fear I have lost all."[42] His immediate reason for pessimism was the address given by Sir William Thomson to the British Association at Glasgow in September, during which Thomson had explained and demonstrated (though without success) a telephone given to him by Bell. Thomson described the experiments he had witnessed at the Philadelphia Exposition in July, and summed up for his audience the essence of Bell's as-yet-unpatented invention: "the mathematical conception that, if electricity is to convey all the delicacies of quality which distinguished articulate speech, the strength of its current must vary continuously and as nearly as may be in proportion to the velocity of a particle of air engaged in constituting the sound."[43] Even before Thomson's speech, the journal *English Mechanic* had described another early Bell device: a telephone that used a stretched membrane "like goldbeater's skin" to pick up sound vibrations.[44]

The implications of Thomson's Glasgow disclosure would later become a major question in litigation, but in the shorter term Bell relied on disclaimer proceedings to salvage his rights. This course of action, equivalent to reissue in the United States, involved applying to a commissioner for permission to remove or correct invalid claims in an already-issued patent. As with reissues, disclaimers were a highly useful tool for tailoring a patent to a particular litigation strategy; especially when, as in Bell's case, a specific threat to the patent had been detected. Bell's disclaimer proceedings, completed in February 1878, renounced four of the six plans and cut the number of claims from eighteen to eight. Multiple telegraphy devices and the autograph device fell by the wayside, along with claims for the sounding box, speaking tube, and the means of adjusting the positions of the sounding plate and electromagnet. What remained comprised most of the claims for the speaking telephone, along with a new paragraph explicitly disclaiming the goldbeater's-skin instrument that had been described in the *English Mechanic*.[45]

The net result of the anticipations and disclaimer was a much-narrowed patent: not so much because of what Bell explicitly disavowed (the *English Mechanic* receiver) or even what he now omitted from his claims (the multiplex telegraph) but ultimately because the grant could no longer purport to lay claim to the broad undulatory current principle. The disclaimed specification did not concede this point, and a final determination still depended on litigation. But the general claims of Bell's patent were most likely now compromised, leaving reliable protection only for a specific device, the magneto telephone. This invention continued to be an essential part of telephone receiver technology in the coming years, but elsewhere left a vacuum into which other powerful patents could flow.

Above all, the absence of a totalizing claim for Bell opened up control over the second generation of telephone technology: the variable-resistance transmitter. Bell's practice of using magneto instruments for both transmission and reception gave way almost immediately to the superior transmitters of Thomas Edison and Francis Blake, which delivered the sound quality necessary for telephony to become a commercial proposition. The variable-resistance method on which these instruments were based appeared in several early forms. One (largely impractical) variant involved a wire moving into and out of a conducting liquid under the action of the voice; both Bell and Elisha Gray described this arrangement in their first telephone applications. A second type operated on the principle that the resistance between two lightly touching or compressible contacts changed as they were pressed together. This technique appeared with the American patent applications of Emile Berliner and Thomas Edison in 1877 and received theoretical explanation—and the name "microphone"—from the physicist David Hughes in 1878. Although the variable-resistance type became ubiquitous in commercial telephony through Edison's and Blake's carbon transmitters, the Bell Company had no need to enforce either patent as long as Bell's undulatory current claim controlled every type of telephone. In Britain, by contrast, Bell's failure to assert an overarching claim made the existence of improved methods for producing electrical undulations highly pertinent.

Because Emile Berliner did not apply for a patent in Britain, Thomas Edison became the first to seek protection for a practical variable-resistance transmitter. Edison obtained two major telephone patents in Britain, one considerably more important than the other from a legal point of view. The first, filed in July 1877, described various ways in which a diaphragm might vibrate a conductor against a "tension regulator"—another conductor whose resistance changed under pressure.[46] This grant established Edison's priority in variable-resistance transmission by variation of contact and became the foundation of his legal position in Britain. The same specification also laid out the basic principle of the electromotograph receiver, Edison's invented-to-order solution for avoiding Bell's rights over the magneto receiver. Instead of using magnetism, this ingenious but slightly impractical device (it had to be constantly cranked to work) generated sound from the changes in friction that occurred between a contact and a moving surface treated with electrolytes when a changing current passed between them. Both the transmitter and the receiver were further refined in Edison's second patent, filed in June 1878.[47] The new specification introduced the carbon-button telephone, a superior application of the variable-resistance principle that used a lump of compressed carbon powder as the tension regulator, and laid out an improved version of the electromotograph. This second grant essentially described the instrument with which the inventor's British company entered business.

Like Bell, Edison combined his American patents for the purpose of making applications in Britain, but Edison's awesome productivity during the summer of 1877 made the relationship between individual American and British patents less straightforward than in Bell's case. Edison's first and second British telephone patents contained thirty and thirty-nine claims, respectively, and combined aspects of at least a dozen American specifications between them. Each of the British grants contained multiple inventions, bringing together designs for transmitters, receivers, and recorders. The application of July 1877, for example, included a design for sound recording alongside the telephone, making it the first British phonograph patent as well as a fundamental telephone grant. Amid this tangle of merging specifications, the language of claims was not usually taken verbatim from American patents. In the 1877 patent, only six out of thirty claims were shared with the equivalent U.S. applications.

Also like Bell, though, Edison found his patent shaped by events beyond the initial drafting process. Three disclaimers over the course of the next five years wholly reworked his principal telephone patent. The Edison Telephone Company of London began to pursue the first disclaimer almost immediately after incorporation in August 1879. This move was driven by litigation strategy. Edison's lieutenant Edward Johnson and the board of the Edison Company had nothing less than total legal victory in mind: they aimed to claim the broad principle of variable-resistance transmission and reasoned that "if we win this fight, the Bell Co. will be run out of Europe & we will have the field alone."[48] In preparing for litigation, the company's legal advisors identified several formal weaknesses in the specification that required amendment to put the patent in fighting shape. They disagreed, however, on the best approach to amending its text. Edward Johnson and the eminent patent agent and solicitor J. H. Johnson (no relation) wanted to retain every claim that did not jeopardize the patent. Richard Webster, the firm's counsel and already a rising legal star, advocated a ruthless paring-down. Webster's approach prevailed on the grounds that it offered Edison's opponents the fewest opportunities to plan a counterattack.[49] Edison's disclaimer duly issued in February 1880. Of the thirty original claims, four remained: the use of a mica diaphragm, the combination of a diaphragm and tension regulator, the electromotograph receiver, and the phonograph.[50]

The Edison interests' aggressive strategy proved in vain, however. Edison's patent reached litigation readiness too late to save his company as an independent concern. By the time the disclaimer issued, the Bell firm's financial advantage had overtaken Edison's legal preparations and shortly afterward forced the Edison Company to terms.[51] The benefits of Edison's reworked patent thus fell to the merged United Telephone Company, which performed two further amendments. The first of these, in June 1881, removed the electromotograph claim, now surplus to requirements since the United Company owned Bell's superior magneto receiver.[52] The second disclaimer had a more obvious defensive origin. During litigation in 1882, the Edison patent as a whole was temporarily held invalid because the phonograph portion had not appeared in the "provisional specification" submitted in advance of the full patent application. In response to this finding the United Company stripped out the phonograph claim and for good measure dispensed with the mica diaphragm claim, which had survived a minor challenge to its wording.[53] At the end of the process, the Edison patent had been reduced to a single claim—the combination of a diaphragm and tension regulator—that bore the full burden of delivering broad control over the variable-resistance telephone transmitter.

Far from simply replicating Bell's and Edison's intellectual property in Britain, the process of international transfer altered the patents at three points: drafting, disclaimer, and context. In the drafting process, claim terms were apparently chosen primarily for their precision in describing the invention, rather than with an eye to legal strategy. With the disclaimer process, legal strategy came to the fore. The patentees focused on eliminating potential vulnerabilities and, in the process, boiled the patents down to their most important elements. Finally, the new context of inventive priority altered the meaning of the patents even where it did not force a change to the text. The timing of prepatent public disclosures meant that Bell's and Edison's British patents possessed a different relationship to each other and to telephone technology in general than did their American prototypes.

The combined result of these refractions was that Bell's British grant covered one type of instrument, indispensable as a receiver, rather than telephony as a whole; Edison's laid claim to another type, soon indispensable as a transmitter. It is tempting to say that the Edison Company, had it survived commercially, might have realized its objective of driving

Bell out of British telephony through total control of the variable-resistance method. This outcome was certainly no less likely than Bell driving out Edison by legal means: neither patent appeared likely to control the other; and of the two, the Bell grant faced the more obvious challenge to its validity.[54] Such projections are, however, of little use in the history of patent litigation. The result of a hypothetical Bell versus Edison lawsuit cannot be deduced from the outcomes each patent achieved against other opponents. Again, what mattered was who sued whom, when, and where.

§

The United Telephone Company established its broad patent coverage in a series of cases tried between 1882 and 1885. For the company, these lawsuits were test cases, prompted in large part by the need to establish the patentees' rights definitively and publicly. The effort was, despite one brief setback, wholly successful. Like the *Spencer* and *Dolbear* decisions in the United States, the judgments in the United Company's test cases construed the basic patents to control every available form of telephone. Because the British patents represented quite different raw material, however, the suits necessarily followed a different course. Edison's patent assumed much of the burden of monopoly, bringing with it a set of scientific uncertainties that hardly figured in the American trials. Meanwhile the questions surrounding Bell's grant centered more on his botched international transfer than on his priority per se. Each of these challenges to the United Company's patent position was further shaped by the particular adversarial contests that commercial telephone competition produced.

As the legal campaign began, two main issues stood out. The first concerned the possible invalidation of Bell's patent by prior publication. This charge was not quite analogous to the claims of prior invention that proliferated in the United States, since British law gave no credit for discoveries made in obscurity. Instead the test was whether the telephone invention had been published in Britain before the date of Bell's patent. Thus Reis's instrument could overturn Bell's rights if, and only if, knowledge of his invention had been available to the British public. A more obvious threat than Reis, however, came from publications of

Bell's own designs during the period of delay in obtaining protection in Britain. Sir William Thomson's demonstration of the telephone at Glasgow in September 1876 was, to the knight-scientist's great embarrassment, the greatest danger to the validity of Bell's patent.[55] To save the grant, the United Company had to exploit to the full its one piece of good luck: that the Glasgow telephone had failed to work when presented. Unbeknownst to Thomson, the vibrating steel disc of the receiver had been screwed down for safekeeping during the transatlantic crossing and had then been accidentally bent upward to form an arrangement like an open tin can. Thomson had assumed this configuration to be the correct operating mode, when in fact the disc should have lain flat and unfastened on top of the cylindrical sounding box. Unable to elicit speech, Thomson had shown off the bent-disc version as a silent exhibit at the British Association meeting. Just as importantly, the distorted instrument had appeared in reports and illustrations of the event. Based on this flawed version of prior publication, the United Company could argue that no disclosure of Bell's invention had been made.

The second major area of uncertainty was whether Edison's rights covered the expanding range of telephone transmitters. Although modern technical histories routinely classify early variable-resistance transmitters as "microphones," the term was a subject of legal and scientific controversy in Britain during the late 1870s and early 1880s. The first public exposition of microphone theory came from the Welsh-born, Kentucky-based telegraph inventor David Hughes, in a paper read to the Royal Society in May 1878. Hughes described an arrangement of three nails, two forming an incomplete circuit and the other lying across them in an "H" formation, and showed that light contact and a small contact area allowed for substantial changes of electrical resistance in response to small vibrations of the surrounding air. Another variant used a carbon "pencil" with sharpened points at each end resting loosely in indented carbon blocks. Such was the sensitivity of the device—which popularizers declared could capture "the tramp of a fly"—that its inventor likened it to an acoustic version of the microscope, hence "microphone."[56] Hughes's refusal to patent his discovery won scientific plaudits at a time when zealous electrical patenting had begun to raise concerns about "speculation."[57] It also created a widely held im-

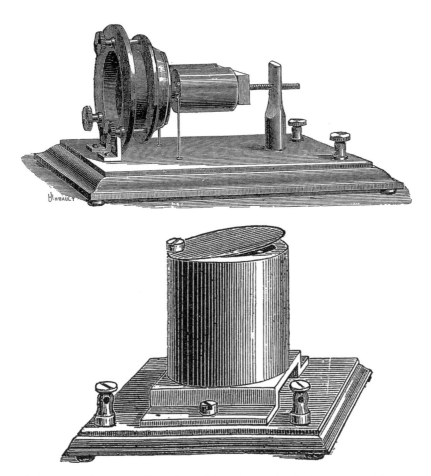

The "Glasgow Instrument" nearly undid Bell's British telephone patent. Sir William Thomson displayed it at the British Association for the Advancement of Science before Bell's application had been filed. Fortunately for Bell, the steel disc atop the receiver had mistakenly been left screwed down, and the device would not work. The courts decided that it was not a prior disclosure of the invention. "Bell's Articulating Telephone," *Engineering*, December 22, 1876, 518.

pression among British electricians that the microphone principle lay in the public domain.

Thomas Edison had other ideas. Not only did he insist that the Hughes paper recapitulated the principle of his own British patent, which had issued in October 1877, but he furiously accused the Post Office's chief electrician William Preece of betraying his confidence to Hughes.[58] This attack on a leading figure of British electrical science did nothing to dispose London editorialists favorably toward Edison and his claim. More pressingly for the owners of the Edison patent, Hughes's disclosure of a potentially unpatented transmitter immediately began to shape the technical choices of would-be market entrants. Inventors of new variable-resistance transmitters presented themselves as followers of Hughes, describing their devices as microphones and in many cases building directly on the carbon-pencil design.[59] The United Telephone Company responded to the appearance of the first commercial carbon-pencil transmitter by buying the patent for £20,000.[60] More buyouts of this order would quickly have proven unsustainable. Facing widely expressed doubt about whether these microphones fell within Edison's rights, the United Telephone Company had powerful incentives to seek a legal resolution in its favor.

The litigation began with a pair of test cases in 1882, *United Telephone Co. v. Maclean* in Edinburgh and *United Telephone Co. v. Harrison, Cox-Walker & Co.* in London. As the first telephone patent trials to approach a decision on the merits, both cases received rapt attention from the electrical press, which reprinted page after page of expert testimony and cross-examination in full.[61] Both were test cases carefully chosen by the United Company.

The timing of the test cases reflected the evolution of the United Telephone Company's competitive priorities. From its formation in May 1880 until the later part of 1881, the firm did not attach great importance to the need for a ruling on the telephone patents and typically settled infringement suits out of court. The atmosphere of impending patent warfare that had gripped the Edison enterprise during its confrontation with Bell had dissipated with the merger of the two groups and faded further into the background while the United Company concentrated on fighting off Post Office control. Only when the company emerged from its settlement with the Post Office in the spring of 1881 did legal action on the patents begin to play a role in clearing the field of infringers and paving the way for provincial expansion. At this stage, however, the United Company's policy was one of aggressive acquisition backed up by litigation. When the Paris-based American Frederic Gower attempted to float a rival telephone company in Britain, the United Company disrupted the flotation through patent threats and then absorbed the Gower interests by forming a jointly owned manufacturing company. As the patent-holding group expanded geographically by promoting regional operating companies, lawsuits were brought against independent exchange operations such as D. & G. Graham in Glasgow and Moseley and Sons in Manchester. By the end of the year, both had agreed to sell out to the incoming United Company affiliates.[62] Acquisition successfully removed the most pressing competitors, but did so without producing a court decision on the patent rights that could effectively discourage other entrants. By late 1881 the costs of this strategy had begun to rise, in the form of spreading infringements and general uncertainty about the status and scope of the patents. The United Telephone Company was now "desirous of getting a verdict."[63]

Industry observers found it a "matter for comment" that the patentees sought a Scottish decision first rather than pursuing the growing crop of infringements in London.[64] The advantages of the strategy soon became clear. Alexander Maclean was a small-scale infringer, a telegraph engineer who had installed telephones for a firm of solicitors.[65] His instruments, supplied by Messrs. Theiler and Sons of London, were built on the Hughes carbon-pencil model and presented a suitable test of whether Edison's patent covered the growing field of so-called microphone transmitters. Maclean's most important attribute was his Edinburgh location. Although a Scottish case could not produce binding precedent in English courts, it maximized the prestige value of the company's star witness, Sir William Thomson, who appeared in this case not only as Scotland's leading electrical authority but also as a key participant in the events under dispute.[66] Thomson's premature display of the Bell device in Glasgow presented major difficulties for the United Company. By putting Thomson on the stand to explain that he could never have produced a working telephone from the information available to him at Glasgow, the company could rebut the suggestion that anyone else could have done so.

Thomson's presence was only one example of the way in which—quite apart from the merits—the United Company held the whip hand in the Maclean trial. Scotland's two foremost advocates argued the firm's case, one of them the lord advocate (the government's minister for legal affairs in Scotland) and the other the Scottish solicitor general. In expert witnesses too, the United Company had a substantial advantage. Along with Thomson it retained the electrical engineer Fleeming Jenkin, fresh from winning the highest award of the Royal Society of Edinburgh, and two leading consulting engineers and "professional witnesses" of the first rank, Sir Frederick Bramwell and John Imray. Meanwhile Maclean's expert witness was Conrad Cooke, science correspondent of the journal *Engineering*. Cooke possessed a modicum of influence in the scientific community, which he had exerted against Edison ever since a brief association with the Edison Telephone Company had ended acrimoniously.[67] But he hardly balanced the scales.

The United Company's overwhelming firepower was amply reflected in the judgment of Lord McLaren, delivered on February 1, 1882. McLaren "accept[ed] in all its particulars" Bramwell's technical description of the telephone, and with it the certainty of an inventive gulf between Reis's make-and-break discovery and Bell's undulatory current.[68] From Thomson, the judge accepted that the Glasgow speech had aimed "to describe results rather than processes," and had resulted in too incomplete a description of the telephone to constitute prior publication.[69] Turning to Edison's rights, Lord McLaren rejected Maclean's first defense, that the patent was invalid because of a discrepancy between the provisional and complete specifications. The microphone issue, being the most technically complex part of the case, McLaren found "a question of some difficulty although I have ultimately formed a clear opinion."[70] Maclean's lawyers made a subtle argument in asserting that the microphone differed from Edison's design. They proposed that Hughes had achieved variable resistance from the delicate contact between two conductors, while Edison had believed it to stem from the compression of a semiconducting material—hence his specification of an elastic substance for the "tension regulator" such as silk fiber rubbed with carbon. Describing Edison's patent, counsel for Maclean announced that "the principle running through all his descriptions is compression."[71]

This claim struck at the heart of expert uncertainty about the variable-resistance principle. Contemporary understanding of the instruments simply did not provide any clear answer about what went on at the point of contact.[72] In a situation of indeterminate science, however, Lord McLaren again deferred to the United Company's experts. Whatever the molecular interactions taking place, he declared, Edison's transmitter and the microphone necessarily acted by the same method.[73] The judge therefore concluded by awarding Edison his broad construction. Edison's patent was emphatically "not for a specific instrument," and could embrace any variable-resistance device.[74]

The case concluded with both the United Company and its critics looking ahead to the next phase of litigation. Although the *Maclean* decision would not bind an English court, the company's solicitors formally requested a certificate of the patents' validity to brandish in English proceedings.[75] Meanwhile the editors of the *Electrical Review,* summing up the case for their readers, predicted a sterner test of the patents in the London courts, "both sides being armed with a more formidable array of capable and personally-disinterested electricians than appeared during the late trial."[76]

§

That test began in late April with opening arguments in *United Telephone Co. v. Harrison, Cox-Walker & Co.* The case revisited all the questions raised in *Maclean,* but now with two additional challenges for the United Company. First, the defendants represented more substantial opposition. Although Harrison, Cox-Walker & Co., an electrical engineering partnership based in Darlington, was the named defendant, the primary contestant of the case was the London and Globe Telephone Company, a largely American-run venture then making a serious attempt to enter the exchange business in London.[77] The London and Globe Company was not particularly well financed when the suit began, but it was determined and patent savvy, with a portfolio of British and European patents assigned by reputable inventors such as Amos Dolbear and Thomas Watson.[78] One London and Globe acquisition in particular—a patent purchased from the Yorkshire clergyman Henry Hunnings— presented the United Company with its second challenge: securing control over yet another distinct type of variable-resistance transmitter.[79]

Instead of using single contacts like the Blake and Edison devices or carbon pencils like the Hughes-type microphones, the Hunnings transmitter employed loose carbon granules between metal contacts. The large number of electrical contacts made within the pile of granules gave the device superior performance—so superior, in fact, that the carbon-granule type became the standard transmitter in both Britain and the United States from the 1890s onward.[80] Even in its early and imperfect form, Hunnings's device represented an important advance that the United Company could hardly afford not to control and, more importantly, a viable form of transmitter that might fall entirely outside the patent. The marked physical dissimilarity between the Hunnings transmitter and Edison's invention made United Company's legal task harder. In particular, the question of infringement turned on whether or not the Hunnings transmitter possessed a "diaphragm" (also referred to as a "tympan"). Edison's patent had been so stripped down by disclaimers that by this point its only relevant claim was for the combination of a diaphragm and tension regulator in an electric telephone. The court in *Maclean* had assumed these features to be essential elements of Edison's broader invention, that is, variable resistance by varying contact. But Hughes's experiments had shown what Edison had not realized: that signals would be produced by sound vibrations operating directly on the conductors. If a working instrument such as the Hunnings transmitter could be shown to operate without a diaphragm, Edison's patent would lose its stranglehold on the variable-resistance principle.

Arguments at trial focused primarily on the semantic issue of what constituted a diaphragm. Counsel for the United Company argued that the thin piece of platina foil on one side of Hunnings's carbon granules amounted to a diaphragm. Witnesses for Harrison, Cox-Walker & Co. responded that it was nothing more than an electrode. Further hairsplitting accompanied the second-most debated question, the status of the Glasgow anticipation. Less attention was devoted to the array of other defenses, such as the argument that Edison's patent showed an impermissible difference between the provisional and final specifications, or the supposed anticipation by Reis—although the United Company did take the trouble to deny that Reis's publication in a German-language journal could constitute public disclosure in Britain. Both sides mobilized significant scientific manpower in support of their positions. The distribution of expert testimony was considerably more equal than it had been in Edinburgh, although the parties called rather different types of witness. Again, alongside Sir William Thomson, the United Company put forward the cream of the expert-witness profession: Sir Frederick Bramwell, Dr. John Hopkinson, and John Imray. Their opponents meanwhile relied on academic electricians such as Silvanus P. Thompson, William Fletcher Barrett, and William Ayrton, along with the American inventors Amos Dolbear and Thomas Watson, both of whom had connections with the London and Globe Company. Legal talent was also more evenly matched, though again, as with the cadre of witnesses, the leading specialists in patent litigation appeared for the United Company. Richard Webster, Theo Aston, H. H. Cozens-Hardy, and John Fletcher Moulton all represented the monopoly; lesser lights (though still experienced patent lawyers) appeared for Harrison, Cox-Walker & Co.

In deciding between the competing interpretations on offer, Mr. Justice Fry took the United Company's side on almost all counts. The Glasgow demonstration's inadequate disclosure, he ruled, was ultimately "proved . . . by the fact that it failed in the hands of perhaps the most skilled person in the world to deal with such an instrument—Sir W. Thomson himself."[81] Reis's instrument may have been "published" in Britain, since one Dr. Muirhead had apparently read the German article in the library of the Institute of Civil Engineers, but its make-and-break working failed to anticipate Bell. On the question of Hunnings's infringement, Fry allowed that Hughes's diaphragm-free experiments "suggested the argument that Hunnings's and Edison's transmitters operate on different principles." But he pushed this possibility aside and in doing so preserved the Edison broad claim, by reading a diaphragm into the Hunnings specification. Defining a diaphragm as any partition across which vibrations would carry, the judge found Hunnings's platina foil to fit the description.[82]

This construction of the Edison patent was broad enough to cover all existing commercial transmitters. Justice Fry's decision did contain a nasty surprise for the patentees: the judge reluctantly explained that the Edison grant was invalid because of an inconsistency between the inventor's provisional and final patent applications.[83] Fry all but apologized for ruling against Edison, remarking that "however great my sense of doubt may be at the ultimate decision of the case, I must bear in mind that I am a competent judge to decide a case."[84] For the United Telephone Company, however, the decision was only a temporary setback; the patent could be restored by disclaiming its offending phonograph portion. Once this disclaimer had been achieved in August 1882, Fry's broad construction of the grant became fully operational. In the meantime, the Court of Appeal also upheld an injunction against Harrison, Cox-Walker & Co. under the Bell patent, distinguishing Bell's receiver from that wholly unrelated instrument, the "Glasgow uncorrected receiver."[85]

The supremacy of the Edison patent was cemented by *United Telephone Co. v. Bassano & Slater*. This case, argued during the summer of 1885, involved an attempt to invent around the Edison patent by creating a telephone with no trace of a diaphragm. Messrs. Bassano and Slater sold a device in which the mouthpiece opened directly onto a carbon pencil, housed within a wooden casing. Because this instrument lacked any sort of partition between the tension regulator and the source of sound, the two electricians from Derby hoped to avoid judgment under the Hunnings precedent. In response, the United Company once again mobilized its expert witnesses, this time to argue that the vibrating back of the wooden box clearly functioned as a diaphragm under the patent. The back board vibrated, Thomson, Bramwell, and the rest recited dutifully, and in doing so passed on its vibrations to the tension regulator; ergo, it was a diaphragm. Despite the incredulous observations of counsel for the defense that very little would *not* be considered a diaphragm under this construction, Mr. Justice North endorsed the theory and found for the United Company.[86] The following year saw the verdict affirmed by the Court of Appeal.[87]

The *Bassano* decision raised a general cry of indignation from the leading electrical and engineering journals, whose editors and correspondents raised their usual tone of heavy-handed sarcasm to a new level. One asked whether a microphone laid on the ground would cause the latter to become a tympan—and, if so, had Edison intended to patent the Earth?[88] Notwithstanding the hail of criticism directed specifically at Justice North, two important points about the judgment bear making. First, the logic of North's apparently absurd decision already existed in Justice Fry's earlier ruling. Like North, Fry had construed Edison's rights *despite* rather than according to the text of the patent. Edison believed at the time of his patent that the diaphragm performed a necessary function in the variable-resistance transmitter. It did not. To prevent this fact from restricting Edison's rights, Justice Fry had defined the diaphragm according to a set of descriptive characteristics rather than any active function. Rather than requiring plaintiffs to show that the alleged diaphragm played a role equivalent to Edison's, Fry had shifted the burden onto defendants to show that they did not have any component that could possibly be regarded as a diaphragm. Justice North merely confirmed that this test was impossible to pass.

Although North did not materially expand the patent monopoly, he certainly drew attention to it. Thus the second important point about the *Bassano* decision emerges: that condemnation of the ruling merged into a broader allegation that the courts twisted scientific evidence and patent specifications in order to produce broad rights. The telephone became one of three chief exhibits, along with Edison's electric lamp and the Otto gas engine, in a charge of judicial "patent warping" leveled by Silvanus Thompson in the pages of the *Times* and much echoed across the engineering press.[89] Edison's lamp litigation, in particular, replicated the pattern of the telephone suits in the second half of the 1880s.[90] In that matter, the question of infringement depended on the definition of a "filament," which the courts interpreted broadly to cover the quite different shapes and materials used by competing inventors. As the *Electrical Review* complained, "Everything is really either a filament or a diaphragm."[91]

In all these cases, the connection between monopoly power, legal and scientific resources, and broad patent coverage delivered by the courts was lost on no one. The *Times* found Thompson's "protest against the seductive influence of experts and crafty counsel, though somewhat superfluously animated . . . not wholly inopportune."[92] Editors of the *Electrical Review,* fixing on the role of expert witnesses in the telephone decisions, wondered when the patent law would be free of the "shifting and evasive assertions of partisan scientists, who quibble over gnat-like niceties, and, when it suits their purpose, complacently close their eyes whilst a whole camel is gulped down."[93] The *Pall Mall Gazette* meanwhile saw *Bassano* as part of a persistent and "mischievous practice of enlarging in favor of a powerful monopoly the claims of a patentee."[94] If the critics were to be believed, science in the hands of the lawyers had come dangerously adrift from science in the eyes of the engineers, and legal monopolies pursued through the courts were increasingly aligned with economic ones.

§

With hindsight, it is possible to see that the various critiques of monopolistic patents in the 1880s converged on a common point: judicial favor for "pioneer" grants. As Silvanus Thompson pointed out, the prerequisite for a patentee receiving the "judicial benefit of having his patent stretched" was his acceptance by the court as the "great and original inventor of a fundamental invention of absolute novelty."[95] In other words, the courts responded to arguments about the pioneering nature of particular inventions by granting extremely broad rights.

The language and logic of the key telephone decisions abundantly illustrate Thompson's point. In upholding Bell's claim against the Glasgow anticipation, Mr. Justice Fry emphasized the inventor's "remarkable discovery"; Sir George Jessel, in the appeal, hailed it as "one of the most important and remarkable inventions of the century."[96] These comments were not empty praise: both judges considered this achievement relevant to their construction of the specification. Fry deemed it "the duty of a judge to give wide interpretation to the words of the document" in such a situation, and Jessel declared during arguments that "a great inventor is not to be defeated lightly on mere technicalities."[97] As well as helping to save Bell from defeat, pioneer status evidently counted for Edison as well. Richard Webster, arguing the United Company's case for a broadly construed Edison patent, stated that "[t]his specification of Edison's came upon the world as a complete revelation, which absolutely started knowledge afresh . . . [and] had the effect of revolutionising the world of science on that particular matter."[98] Counsel made it clear that his words were no mere rhetorical ornament; to the contrary, they were "the particular fact on which the whole case depended."[99]

Other contemporary cases show a similar stress on the originality of the invention under scrutiny. Ruling on the Otto gas engine patent at the beginning of the 1880s, Jessel promoted the importance of construing a patent for "a genuine great and important invention" so as to uphold rather than restrict its reach.¹⁰⁰ Similarly, when the Court of Appeal supported a broad patent for Edison's electric lamp in 1887, the *Electrical Review* claimed that "[n]o one who has followed the hearing of this case can fail to be convinced that the statements by the Edison Company's counsel that before the date of Edison's patent such a thing as an incandescent electric lamp was not commercially known, and that almost immediately after that date such lamps were manufactured by hundreds of thousands, had a very important effect upon the minds of the judges."¹⁰¹

Moreover, the courts were prepared to enshrine this type of reasoning in doctrine. A year after *Bassano* and around the same time as the electric lamp case, the Court of Appeal handed down its 1887 decision in *Proctor v. Bennis*, which became the leading explication of a "pioneer" or "master" patent classification.¹⁰² The judgment granted a fundamental patent for a rotary stoking machine—which the court deemed an entirely new device—on the ground that the novelty of the idea was part of the merit of the invention. Effectively, *Proctor v. Bennis* recognized in British law a binary distinction between pioneer inventions and mere improvements—the standard that, as we have seen, was a well-established part of American law. The rule was not presented as a doctrinal innovation out of whole cloth; it referred to a number of precedents going back to 1841.¹⁰³ But it instantly became an influential statement of the law in infringement cases.

The judges' embrace of pioneer patents both before and after *Proctor* seems somewhat at odds with the prevailing historical view of the British courts. To the extent that there is a historical account of patent adjudication during this period, it suggests that British law had a relatively restrictive set of rules for interpreting inventors' rights. The legal scholar W. R. Cornish has identified a "fence-post" approach to patent interpretation, in which judges treated the claims of the patent as setting strict outer limits on the scope of the grant. Leading opinions issued by the House of Lords during the 1870s established the principle that "everything which is not claimed is disclaimed" and held that there should be no equitable leeway granted to a patent because "that which is protected is that which is specified."¹⁰⁴ Noting that the 1883 Act made explicit claims compulsory for the first time, Cornish describes an increasingly confining approach to patent scope during the 1880s and 1890s. In this view, the interpretation of patents became a kind of linguistic sparring: a "game of determining their meaning . . . played by the standards of the best Chancery minds."[105]

This vision of precise and pedantic adjudication is consistent with other accounts of contemporary judging, which have emphasized the growing formalism of the courts during the later nineteenth and early twentieth centuries. Formalism describes a number of related tendencies supposed to have been prevalent at this time, including judges' myopic focus on precedent to the exclusion of policy considerations, and a reluctance to depart from the literal terms of the legal agreements placed before them. As a judicial style, formalism has generally been associated with the powerful ideology of "freedom of contract." This notion, dominant in much late nineteenth-century adjudication, held sacrosanct the bargains struck by freely contracting parties—in this case, the patentee and the public at large.[106] Many pronouncements in patent law took a similar line. Judges sought to read the patent "like any other document," a maxim which both acknowledged and encouraged the close association between patent and contract law.[107] As one patent agent put it, the education of lawyers made them "attach too much importance to words as distinguished from things."[108]

The telephone cases show that formalism and strict interpretation did not reign supreme. Linguistic nicety did play a role, insofar as the definition of the single word "diaphragm" came to be the central question of the infringement cases. But judges clearly tortured the term in order to construe Edison's patent broadly. More to the point, their crucial assumption that pioneer patents should receive special treatment stands at odds with a neutral, contractual approach to patent interpretation. In general, the formalist or text-centric method never held a monopoly over patent adjudication. Even as some doctrinal prescriptions pushed claim theory toward a strict adherence to the text, a countervailing current urged judges to look beyond the precise language used. Judges were instructed by another edict from the Lords to seek the "pith and essence" of the patented device.[109] Courts thus systematically ventured beyond the written document to investigate the nature of the invention at hand. The coexistence of the "fence-post" and "guide-post" methods confuses the question of whether British patent law favored narrow or broad patents. Given that judges had both expansive and restrictive doctrines available, the real issue is why some should have chosen to emphasize broad pioneer grants at the particular historical moment of the telephone and electric lamp cases.

One obvious possibility, given the language that accompanied broad patent rulings, is that judges were systematically swayed by the idea of the heroic inventor. The patent abolition campaign had ebbed by the 1880s, leaving the legacy of its opponents in place: a pantheon of heroic inventors durably implanted in popular attitudes toward technological change.[110] The paradigm of individual genius was thus both culturally ascendant and politically vindicated by the passage of a new patent law. How this affected any given patent is a complex question. The heroic inventor ideal did not always operate to the benefit of a patentee before the courts. At least one later observer suggested that it led the courts to strike down many patents for "want of invention" because judges took famous inventors such as Watt as their benchmark and disparaged smaller, incremental improvements.[111] In addition, the patentee of the late nineteenth century was an imperfect fit with the cultural ideal. Recent works on popular and literary conceptions of invention have pointed out that the great inventor trope was always based largely on eighteenth- and early nineteenth-century figures, and have argued that the ideal of the heroic inventor became increasingly nostalgic and old-fashioned as the Victorian age went on. At least one such account contends that the idea of individual inventorship was increasingly discredited in the 1880s and 1890s by the corporate organization of innovation.[112]

The telephone cases, however, suggest that the heroic idea was very much alive in the British courts. Judges' decisions saw repeated expressions of judicial deference to original genius. Above all, judicial favor descended on Thomas Edison, to a degree that eventually frustrated the ever-skeptical *Electrical Review*. "Edison's name is a sort of fetish," the journal's editors complained, "and the public, including the judges, have come to regard Edison as a sort of phenomenal inventor."[113] Certainly Sir George Jessel, the most voluble judge on the subject of deserving pioneers, was known for his attention to public sentiment and would make a plausible conduit through which widespread acknowledgment of a great inventor might feed into legal success.[114] An alternative model might be Justice North, who struggled with the technical side of patent cases and whose admitted difficulties in following the telephone arguments eventually led him to assert that the world-famous Edison had invented electric telephony.[115]

North's difficulties point to another issue, which may well have been just as important. Pioneer patent rulings resulted from adversarial litigation, and so reflected the capabilities and arguments of lawyers for the parties. More than one observer of the legal scene thought that the power accorded to broad patents in the 1880s was a "tendency . . . deliberately fostered by some eminent leading counsel and by a few professional experts, who lend themselves to this mode of securing a monopoly for the patentee."[116] Much authority in British patent law lay with the repeat-player professionals who represented litigants, rather than with judges. The imbalance of expertise was a function both of the specialized legal niche that patents occupied—dominated by a small number of barristers but mastered by few judges—and of the complex technical questions that the courts were required to adjudicate. William Grove, on the cusp of becoming a judge after distinguishing himself as a patent lawyer, outlined the implications of a disproportionately powerful patent bar to Parliament's select committee on patents in 1871. Among "the inevitable evils resulting from the counsel being superior to the tribunal," Grove worried that courts must either submit meekly to one party's chosen result or rule perversely to demonstrate their independence. Grove tactfully refrained from saying how often such mismatches occurred. Nevertheless, he warned from experience that "you cannot have anything worse than having the Bar very much stronger than the Bench; you cannot produce a greater injustice."[117]

In the years that followed, the rise of the professional witness further augmented the litigant's side. By 1883, a member of parliament could cite the lord chancellor as complaining "that the expert witnesses who appeared in patent trials were absolutely the masters of the Court, the suitors, and the jury."[118] At the technological cutting edge of the second industrial revolution, all these imbalances were exacerbated: the premium on scientific authority in the courtroom was at its highest, the disparity between judges' remit and their abilities was occasionally glaring, and the pool of well-qualified lawyers was small enough to be monopolized by just one of the parties in an action. During the struggle for control of new electrical technologies, in particular, "the more wealthy of the two parties . . . [was] generally found to have retained, by general retainer, all the leading counsel who have distinguished themselves in electrical patent law."[119]

In light of the power wielded by counsel and professional witnesses, the telephone company's occasionally strange legal victories are somewhat easier to understand—as is the depth of controversy over pioneer patents. Patent lawyers might not take sole credit for the judges' responsiveness to claims of heroic invention, but they certainly exploited any such tendency to the full. By doing so, the repeat players of the patent system ensured that legal outcomes reflected not only the law per se, but the configuration of legal practice itself.

CHAPTER 7

Patents, Firms, and Systems

STANDING BEFORE THE JUSTICES of the U.S. Supreme Court in 1887, Bell counsel James J. Storrow recounted the words of a beaten opponent. "It seemed to him," Storrow recalled, "that this whole telephone system was like a pyramid balanced on its apex; that this vast system all over the world to-day was based on this one little imperfect machine in the Bell patent."[1] In the heat of the nineteenth century's biggest patent trial, with competitors poised to rush in should the Bell claim fall, the image of a great monopoly teetering on the narrow foundation of Bell's rights seemed highly apposite. Yet by the time that Storrow spoke, the telephone business rested on an altogether broader base. Since 1880 the American Bell Telephone Company had transformed itself from an overstretched patent franchiser into an enterprise of national scope, overseeing an increasingly coordinated assemblage of affiliated companies and interconnecting networks. This corporate grouping was a powerful incumbent of the telephone field, whose leadership would not easily be challenged even if the patents were wiped out.

In explaining the big picture of industrial change—why great companies rise and fall; why technologies move between "closed" monolithic systems and "open" innovative ecosystems—the role of patents can be surprisingly hard to pinpoint.[2] The telephone provides a striking example. Treatments of the industrial structure of telephony have traditionally paid much more attention to managerial strategies and regulation than to the fundamental patents. There are at least a couple of possible explanations for this emphasis. One is that patents were of limited importance. After all, telephone service proved capable of taking monopoly form without controlling patents; it would eventually do so in the United States, and indeed did so from an early stage in most European countries and in Canada. Rather than the patents, then, the more salient story line might follow other pressures for concentration and consolidation, be they the economics of the industry, public policy, or both. Another possibility is that the patents present a very simple story: they excluded competition, and then they expired. For most writers on American telephony, the significance of the patents is entirely summed up by the transition from a closed market before 1894 ("the era of patent monopoly") to an open one afterward ("the competitive era"). As one representative author explained the matter, "Most industrial organizations . . . conduct their business in a somewhat different manner under competition than under monopoly."[3]

The effects of the fundamental patents were neither simple nor incidental to the way that the telephone industry developed. The patents were not solely defensive instruments, but worked as sinews of an emerging Bell-affiliated group of companies—linking firms, coordinating technical standards, and channeling money, innovation, and government policy into a particular vision of the new technology. Understanding the role of the basic patents means looking beyond the litigious suppression of competition to include the close connection between intellectual property and organizational development. By closing off market entry to competitors, patents became the single most important influence on the early growth of the telephone: the nature of the service, and of the companies that provided it, were different than they would have been otherwise. Even after legal protection had expired, the legacy of patent monopoly lived on in the strategies and competitive positions of the industry's powerful first movers.

To be sure, the patents had not created a single "Bell System." That would come later, in the early twentieth century, when the financial and managerial integration of the Bell companies was much greater. After

1907, Bell executives poured energy into promoting the idea of a centralized, efficient, unified national telephone enterprise—promoting it both internally, to the still-separate local companies whose operations they oversaw, and externally, to a public whose suspicion of bigness and monopoly the telephone giant needed to overcome.[4] To read this carefully constructed image of unity back into the nineteenth century and the earliest years of the telephone business would be a mistake. Nevertheless, there is no doubt that the existence of a patent monopoly gave the new technology an organizational coherence that it would not otherwise have had. In the mid-1890s, there was no obvious reason why telephone operating companies in Texas, California, and Chicago should be linked to a Boston firm, or should share equipment and practices with each other. All of these things were highly unlikely, given the managerial and capital requirements involved: telephone service was, for many reasons, a stubbornly local and decentralized business. Yet the patents produced just this situation.

As will become clear, these arguments implicate many other factors in the corporate integration and competitive structure of telephony. In general, legal monopoly was thoroughly entangled with the other forces that shaped telephone provision—that was, in fact, the reason why the patents mattered so much.

§

In the earliest days of commercial telephony, Alexander Graham Bell outlined his conception of the industry in a letter to his English investors. By way of analogy, he invited them to reflect on the "perfect network of gas-pipes and water-pipes" that ran under every large city, radiating out from a central source in a pattern of main arteries and side pipes that branched off into homes and businesses. "In a similar manner," Bell proposed,

> it is conceivable that cables of telephone wires could be laid underground, or suspended overhead, communicating by branch wires with private dwellings, country houses, shops, manufactories, &c., &c., uniting them through the main cable with a central office where the wire could be connected as desired, establishing direct communication between any two places in the city. Such a plan as this,

though impractical at the present moment, will, I firmly believe, be the outcome of the introduction of the telephone to the public. Not only so, but I believe in the future wires will unite the head offices of the Telephone Company in different cities, and a man in one part of the country may communicate by word of mouth with another in a different place.[5]

Even as Bell wrote, the debut of the switchboard in America had begun to make his prediction of urban telephone exchanges a practical reality.[6] Several years would pass before long-distance communication between towns became commonplace. But Bell's vision was fully borne out in the long term: telephony developed as a network industry or, as the inventor put it in 1878, a "grand system."[7]

Like Alexander Graham Bell, modern writers often use the terms "network" and "system" interchangeably when describing large-scale, distributed technologies such as railways, electrical power, or telephone service.[8] Both words carry the sense of interrelated parts combined to form a "complex unity"; both describe structures that pose complicated problems of design and coordination.[9] Yet some subtle differences exist between the two concepts. "Network" generally refers to the physical web of connections between users or places. These links may distribute a product like gas or water, or they may carry traffic from point to point, as in telecommunications or transport. Either way, the network form presents characteristic choices for the service provider: What routes should be followed? Who should have access? How should the costs of shared plant be distributed among users? As a result of choices like these, the physical configuration of the network itself has a powerful shaping effect on the industry in question. "System" is potentially a broader term, referring to a set of interdependent technologies and operations. Thus a telephone network with its exchanges, lines, terminal instruments, and human operators constitutes a system. But so too do nonnetwork technologies like air traffic control or missile guidance, both of which rely on bringing together information, hardware, and operational control. Although the central features of networks involve linking together distant points or nodes, systems imply more general problems of coordination: how to ensure compatibility between components, how to match business organization to technical requirements, how to maximize efficiency in the operation of the whole structure.

Taken together, the network and system aspects of the telephone business have been the central preoccupation of its historians, and with good reason. More than any other features of the industry, the overlapping imperatives of network and system account for the distinctive economics and politics of the telephone. In particular, they each bear strongly on the tendency of telephone service toward monopoly. On the strength of its network characteristics, the telephone was widely regarded as a "natural monopoly" for much of its history.[10] The basis of this analysis was the simple proposition that a single telephone network could connect all subscribers to one another, whereas two or more competing networks would likely refuse to interconnect, thus forcing each set of customers either to join multiple networks or to have fewer connections. In addition, demand for telephone service was subject to an externality known to economists as a "network effect," whereby the value of service to each user increased as more users joined the system.[11] For many interested parties, these features made the appropriate structure of service quite clear. "The telephone being a natural monopoly," the California Utility Commission stated in 1918, "there should be one universal service, as this will enable complete interchange of communications between all telephone users."[12] Although the patent monopolists may have been the earliest and most fervent proponents of such an argument, their case proved capable of attracting support from users and governments.

Just as important as the telephone's own characteristics, however, was the environment in which the industry emerged. As a network technology, the telephone joined a group of industries that held a special place in the nineteenth-century economy. The great networks stood out because of their infrastructural importance: railways and telegraphs remade the economic geography of nations; gas, water, and tramways transformed the urban environment; later, telephones and electricity began to permeate the office, the factory, and the home. These industries also stood apart from the rest of the economy in an important sense: they posed distinctive problems of economic governance by persistently gravitating toward monopoly. The high fixed and sunk costs of laying track, pipes, and wires made competition between rival networks at once socially inefficient and commercially unstable. Time and again during the nineteenth century, railway, telegraph, gas, and water companies engaged in ruinous price wars that ended only when one competitor absorbed the other to form a capital-bloated monopoly.[13]

Political-economic responses to these failures of competition developed piecemeal. A formal theory of "natural monopoly," based on declining average costs to scale, began to emerge in the 1880s.[14] But the notion that certain industries were congenitally unsuited to competition was already familiar to consumers, investors, and governments. Against this background, public policy played an increasingly active role in prescribing the conditions of network provision. Local, state, and national governments used the means at their disposal—including municipal franchises, statutes of incorporation, and public utility laws—to regulate not only market entry but also prices, quality, and the capitalization of companies. By the late nineteenth century, network industries had become the principal venue for experiments in public ownership and regulation.[15] Telephone service could not escape being drawn into the web of political and economic preconceptions that surrounded similar technologies.

If networks were an outward signal of the kinds of competition and regulation that the telephone industry could expect, then systems provided an equivalent shaping influence on the internal organization of the business. Not only day-to-day operations but also strategic decisions and investments were deeply bound up with system considerations. Routine system management included tasks such as supervising traffic flow, collecting charges, and maintaining the technical integrity of the network. Strategic challenges included negotiating industry-wide technical standards, managing the relationships between interconnecting companies, and pursuing efficiency gains through innovation. Externalities abounded: the failures of one distributor could affect the quality of service provided by connecting exchanges, whereas the adoption of new technology by one firm posed problems of compatibility for others. These coordination problems created their own impetus toward centralization. They encouraged both the horizontal consolidation of connected operations, to form coherent large systems out of fragmented local ones, and the vertical integration of equipment manufacture with service provision, chiefly in support of technical standardization. Telephone systems therefore had a functional justification for monopoly to match that of networks: just as the network form seemed to require single providers in each location, so system requirements appeared to favor unitary management of service from one location to the next.

The significance of system goes beyond organizational determinism, however. As many historians of technology have noted, the *idea* of system is a leading example of the socially constructed abstractions that guide managerial and technological decision making. Thomas Edison, for example, was a "system-builder" who conceived and designed his electric lighting method from generation to transmission to illumination.[16] The individual vision of figures such as Edison, once implemented, laid the foundation for collective perceptions of an industry. For inventors, engineers, and executives alike, "system was not merely a convenient shorthand for physical reality. It was rather one way of interpreting that reality, which emphasized the values of functional hierarchy, operational smoothness, consistency, and central control."[17] During the late nineteenth and early twentieth centuries, these values gained increasing purchase in the management of industrial enterprises. They reached their apotheosis in the large bureaucratic corporation, where they could be thoroughly institutionalized in the form of standards departments, traffic departments, and a dominant ethos of "systems engineering."[18] Each of these stages can be seen in the history of the telephone, culminating in the most famous system metaphor in telephony: the use of the term "Bell System" from the early twentieth century onward to describe the national web of companies affiliated with AT&T.

Historians have dealt with the network and system characteristics of telephony in different ways. One approach, clearest in older histories of the monopoly providers, implicitly accepts the organizational imperatives of unified service and centralized management.[19] This viewpoint was especially compelling in the middle twentieth century, when national telephone monopolies held sway in the United States (under the regulated Bell System) and Europe (under public ownership). These entities embodied—and actively promoted—the view that telephone service was inherently "indivisible."[20]

Most recent historians of the telephone reject such assumptions. In explaining the monopoly structure of the twentieth-century industry, these writers call for greater emphasis on business culture and political choice. What appeared rational or even inevitable to an earlier generation of observers is now presented as the product of power relations and institutional biases. Several developments contributed to the emergence of this new perspective. One catalyst was the breakup of the American and British telephone monopolies in the 1980s, which prompted renewed academic interest in America's period of service competition between 1894 and the 1920s.[21] Other influences are methodological in nature. An ever-expanding literature on the "social construction of technology" has overthrown determinist explanations of technological change and substituted an insistence that technical artifacts reflect the cultural and political values of their producers and users.[22] Meanwhile, a similar critical tradition has flourished in business history, downplaying efficiency-centered explanations of economic outcomes in favor of cultural factors, national institutional endowments, and social and political power.[23] Network industries, being freighted with social significance and deeply entangled with governmental institutions, have attracted much attention from both schools.[24]

Prompted by these insights, historians of the telephone have taken pains to show that the form taken by the industry was chosen and contested, rather than preordained. A number of overlapping arguments have emerged. One strain demonstrates the empirical weakness of monopoly effects in early telephony. By showing that monopoly providers did a poor job of satisfying the diverse markets that existed for the telephone, writers in this vein contend that the model of unified and standardized telephone service was economically quite vulnerable. In addition, they suggest that episodes of competition generated rapid growth in service coverage, meaning that the public interest in a convenient single service was less than clear-cut.[25] Another group of studies examines the ideas and social contexts behind the construction of telephone systems. These works find a variety of ideological support for large-scale unified systems, ranging from Bell engineers' enthusiasm for standards to a broader corporate vision of nationwide communications under the benign hand of AT&T. But they also find a conflicting set of values, often populist and suspicious of monopoly control, that privileged lower-cost, locally managed service. Accordingly, it is argued, the victory of one type of telephone service over another involved an ideological struggle between different visions of the technology.[26] Given the indeterminate economics of service, many of these questions had to be resolved in the political realm. At the local, state, and national levels, governments eventually—and decisively—threw their weight behind monopoly. The politics of networks and systems thus mattered a great deal because they led public policy to foreclose options rather than simply to ratify a shape that the industry had "naturally" taken.[27]

If the overarching theme of modern telephone historiography is that monopolies were made rather than born, then the role of fundamental patent control is ripe for reassessment. The telephone industry might have adopted other forms, but had monopoly imposed on it by legal fiat for the first decade and a half of its commercial existence. Patent ownership allocated market-shaping power to particular actors, and their actions in turn determined the setting within which different groups contested the meaning of telephony. Most importantly, patent control came *first:* it anticipated and contributed to all the other pressures toward concentration.

Basic patents held the telephone industry together before networks and systems had become established. In its early stages, telephone technology was rudimentary and its opportunities imperfectly understood. Network construction proceeded gradually and service remained highly localized. Although regulation at the local level was crucial in shaping many aspects of telephone service, public policy was often indeterminate as to the question of market structure. Patents were, during these early years, the single greatest determinant of both monopoly provision and interfirm relationships. Furthermore, the patents made a fundamental contribution to network and system. At the outset, patent holders were able to exert direct control of other actors in the new industry through licensing and to require certain standards and practices as a condition of participation in the industry. They also began to harden contractual links into financial ones by taking equity in return for licenses. Meanwhile, monopoly returns provided the resources for heavy investment in technology and the construction of long-distance lines, while lack of competitive pressures allowed the monopolists to pursue such investments with an eye to their long-run possibilities.

Finally, patent monopoly laid down a basis for future market structure. The imperative to exploit patent control while it lasted helped to cement financial and corporate ties between the telephone organizations. These in turn fostered technological and managerial integration. As time went on, industry leaders came to apply the system idea to initially scattered operations and privileged those projects that integrated and standardized their operations. In this way, the patent holders sought to capitalize on their first-mover advantage and to entrench their monopoly position beyond the lifetime of their legal rights. They accumulated financial and political power that would later serve as a weapon against market entrants. They also presented policy makers with a de facto monopoly that did much to shape the subsequent debate on the "natural" market structure of telephone service.

In sum, this narrative makes the argument that "history matters"—a point that has gained formal recognition in the study of both technology and industrial organization.[28] In the case of the telephone, patent monopoly began a feedback effect: legal protection subsidized the integration of networks and systems, while networks and systems helped the patent owners to leverage their temporary monopoly into permanent market dominance and further integration of service. By playing a role in the monopolists' accumulation of economic and political power and by having decisive effects on the structure of the early industry, broad patent rights took their place at the center of the telephone's business history.

§

The Bell Company's patent control had two distinct effects on the telephone industry. The "competitive" effect, by which the patents prevented unlicensed firms from entering the market, is by far the better known. Equally important, however, was a "constitutive" effect, through which patent rights governed investments and business relationships within and among firms. Proprietary connections in the form of patent licenses joined the component parts of the telephone industry together when little else did and laid the groundwork for functional and organizational integration.

Alfred Chandler Jr., in his classic business history *The Visible Hand,* offered what is now the most familiar story about patent-using industries in the nineteenth century. In this account, involvement in marketing and distribution was the crucial investment for makers of patented articles. The success of companies such as the Singer sewing-machine enterprise and Cyrus McCormick's harvester business rested heavily on their construction of marketing organizations, which eventually included branch offices and a salaried sales force. Like other companies formed to commercialize new inventions, these firms had begun by selling through licensed agents—a strategy that mitigated capital shortage and allowed rapid progress to market for what was, after all, a time-limited intellectual property asset.[29] As the patent-holding companies gained their feet, however, they discovered the advantages of more integrated sales channels. These included the capacity to offer consumer credit, specialized installation, and after-sales maintenance, all of which were essential in persuading consumers to purchase complex machinery. By the 1880s, centralized sales organizations had become de rigueur for the leading suppliers of new business machines like cash registers and typewriters, as well as for the makers of producer goods like elevators and electrical equipment.[30] Surveying the rise of market leaders such as Singer, Chandler argued that it was investment in organization that enabled such patent start-ups to become permanent business powers. As Chandler rightly observed, "A set of patents without . . . an organization could never assure dominance; an organization, even without patents, could."[31]

Telephone service adopted a model of organization different from Chandler's standard type and, along with it, an alternative role for the controlling patents. Patentees of the new device shared many of the concerns of other machine makers: quality assurance, customer service, market expansion, and so on. But telephony was prevented from creating integrated sales channels by the strongly decentralized basis on which the industry developed. Patent rights were one of very few aspects of the infant industry that operated purely at the national level. Otherwise, the telephone business was decidedly local in nature. Private lines, the first type of telephone service to be marketed, rarely ranged further than a few streets' distance from merchant house to stock exchange or from physician's office to pharmacy. Even after the development of switched networks, the great majority of calls were intra-urban or at most took place between neighboring towns.

From a business standpoint, the advent of the telephone exchange forced providers to put down local roots. The costs of exchange construction lay far beyond the capacity of the parent companies to finance centrally, and so relied on capital being raised in the city or region that would be served.[32] Other kinds of local connections mattered too. Promoters placed much emphasis on connecting each town's social and business elites first, so as to render service attractive to others.[33] Engagement with local politics also proved indispensable because municipal ordinances or franchises were usually required for lines to be erected or cables laid beneath the streets.[34] Given these features, telephony resembled an urban utility industry such as gas, water, or electric lighting far more than it did the long-distance networks of railway or telegraph.

Telephone service thus followed an alternative model of downstream integration, that of the proprietary utility company. In its basic outline, this type involved franchise-holding local companies linked to a parent firm—usually an equipment manufacturer—by exclusive supply and patent agreements. At the center of the relationship lay the patent license contract, from which local investors drew the hope of a monopoly position and the patent-holding company gained a web of tied firms. Proprietary groupings became a distinctive form in the new electrical industries of the 1880s and 1890s. Electric light, power, and traction enterprises adopted the structure for essentially the same reasons as telephony. Generating stations, like telephone exchange companies, required local promotion, whereas development of the new industry required the patentees to involve themselves with service. As one historian of General Electric has put it, the solution was for manufacturers of the new technologies to "invent" their customers in the form of utility companies that they themselves helped to promote.[35]

The proprietary groups that resulted defy easy description. Relations between the constituent companies fell somewhere between arm's-length dealings and managerial integration. Following a practice used by the arc light manufacturer Charles Brush as early as 1879, patent-holding firms took between a quarter and a half of local company stock as payment for licenses, and sometimes accepted more stock as payment for equipment. These shares gave parent firms a direct financial stake in, and a measure of control over, the supply of service. Control only went so far, however: relationships between companies under the patent umbrella were usually more cooperative than hierarchical; technical and managerial innovation moved both up and down the chain of supply, as well as sideways among associated firms. Rather than standing to one side of business organization, intellectual property played a major role in the ongoing management of the structure.

In the United States, the marketing of telephones evolved through a number of steps as the parent firm cautiously adjusted its licensing arrangements to changing conditions. The first licenses granted territorial agencies of widely varying size. Adjustment to local utility promotion began in 1879, when Bell Company managers renegotiated the original agency contracts to create city-specific franchises more suitable to the new exchange format.[36] With this policy the Bell Company adopted a wait-and-see approach. Licenses were to last five years, after which time the patent-holding firm had the right to buy out its licensees for the cost price of their plant. Shortly afterward, though, the pressures of competition with Western Union forced the parent firm to become more deeply involved in exchange business. As Western Union threatened to overrun Bell operations in the biggest cities, the Bell Company moved to prop up key licensees. Affiliates in Boston and New York were bought out, and Gardiner Hubbard poured over $40,000 directly into exchange operations in Chicago. Partly in reaction against Hubbard's almost fatally costly stand at Chicago, the conservative financiers then gaining control of the Bell enterprise refused to embrace further purchases of licensee stock.[37] Only with the end of competition and the capitalization of American Bell did the question of forward integration gradually reemerge.

Assumption of an ownership stake as a matter of clear policy began in 1881, with a second general revision of licensee relations. This time, American Bell undertook to replace the prevalent five-year contracts with permanent licenses. Under the purchase clause of the five-year contracts, buying out the exchange business and operating it directly was theoretically an option for the Bell interests. Such was the premium on American Bell stock in the aftermath of Western Union's withdrawal that one historian of telephone finance has suggested direct operation was feasible, and at least one telephone executive claimed decades later that it had been seriously considered.[38] American Bell's public line at the time, however, never wavered from a commitment to decentralization: in the words of the 1883 annual report, "It has always been our policy to keep local capital and influence in the business as far as possible."[39] The new permanent licenses that gave effect to this policy brought between 30 and 50 percent of each licensee's capital stock—usually 35 percent—to the Bell Company in return for a perpetual franchise.

The new financial constitution that American Bell drew up for the industry after 1881 derived only in small part from a Chandlerian calculation about the benefits of functional integration. By this point the company's leaders had developed a discernible interest in the greater coordination of telephone services, principally in order to maintain quality and to promote the connection of neighboring exchanges.[40] But the acquisition of "franchise stock" served more to extract value from the patent rights than it did to extend meaningful central control. From the parent firm's point of view, rights-for-stock allowed the company to "obtain a permanent vested interest in the telephone business independent of its royalty upon telephones."[41] Permanent franchising extended the horizon of earnings beyond the expiry of the basic grant, as well as adding a second revenue stream in the form of future dividends on operating company stock.[42] Executives in Boston maintained the emphasis on decentralized management, albeit supplemented by greater information gathering. Operating companies, having gained their primary objective in the form of permanent licenses, received no undertakings from Bell other than the right to operate under the patents.[43]

The exchange of rights for stock did not, in itself, commit American Bell to a strategy of deeper integration. Patent holders in other proprietary utility industries took a variety of approaches to the ongoing management of their tied firms. The arc light manufacturer Brush, the United States Electric Light Company, and the American Electric and Illuminating Company preferred to sell off their stakes in central station companies once the utilities were established. Electrical manufacturer Thomson-Houston, which began to organize its own utility companies after 1885, created a separate holding company to take possession of their securities. Among all the lighting firms, the Edison incandescent lighting enterprise kept its licensees closest. They reciprocated by pressuring the parent firm to prosecute its patents ruthlessly against competitors in the field. The Edison group was also the most cohesive, forming a community of interests and technical exchanges under a trade body known as the Association of Edison Illuminating Companies.[44]

Even compared with the Edison lighting enterprise, American Bell had a singular interest in integration with its operating companies. From its initial platform of unpaid-for franchise stock, American Bell continued to advance its ownership stake in the licensees during the patent period. Financial and organizational churn among the operating companies provided the main opportunity for doing so. As the practical range of transmission technology improved and interexchange "toll lines" began to connect local systems with one another, waves of consolidation swept over the industry. Some combinations reached considerable scale without majority Bell ownership, the most prominent being the handiwork of the "Lowell Syndicate," which built a substantial organization in Massachusetts, Vermont, Maine, and New Hampshire between 1879 and 1883.[45] However, American Bell also put its weight firmly behind the consolidation movement, declaring that "[t]he connection of many towns together . . . made it of importance to bring as large areas as possible under one management."[46] In New York State and the Mid-Atlantic region, the result was a set of firms that each encompassed multiple cities and towns. Further west and south, the geographical scale was much larger: entire multistate regions were corralled into territorial companies.[47] From 185 agencies in 1880, the map of Bell telephone service was redrawn to produce 50 companies by 1888 and 35 by 1893.[48]

Mergers characteristically upgraded the parent company's stake decisively. In one instance, the New England Telephone & Telegraph Company, incorporated in October 1883, brought together eight companies, of which Bell owned a 30 percent stake in two and part of an unincorporated third—holdings altogether worth some $670,000. In addition to exchanging these holdings for shares in the new enterprise, the patent-holding firm received $4,268,000 par value of newly created franchise stock. When the dust settled on amalgamation, American Bell possessed $6,215,600 in par value common stock, approximately 60 percent of the New England Company.[49] While New England Telephone stood out as an especially large merger in Bell's heartland, the pattern of advancing franchise stock holdings during consolidation recurred elsewhere, and the parent company further bolstered its holdings by purchasing shares whenever operating companies made fresh issues. By December 1885 American Bell held $21.2 million of shares in its licensees: $13 million in stock received for patent rights and some $8 million purchased in cash, most of the latter acquired in 1883 and 1884. Over the patent period as a whole, American Bell received somewhere between $14 and $16 million in securities exchanged for patent rights, plus more than $20 million in stock purchased.[50] Across the board, the parent company owned about 45 percent of all operating company stock throughout the later 1880s and the 1890s. Although levels of Bell ownership in individual firms varied substantially, a later study by the Federal Communications Commission concluded that Bell had established equity control of "most of the principal licensees" by the expiry of its legal monopoly.[51]

The exchange of patent rights for stock and the consolidations of the mid-1880s created what Bell historian Robert Garnet has called the "foundations for the modern Bell System."[52] This description is slightly anachronistic—the Bell System of the twentieth century was still a long way off—but a reasonable one, since the horizontal structure of service companies under a stock-holding parent firm would remain in place, with regular boundary changes, until the breakup of AT&T in the 1980s. It is also fair to say that the existence of a patent monopoly massively promoted the creation of this corporate form. The $14–16 million worth of franchise stock received by American Bell effectively provided a subsidy to the financial integration of the telephone industry. And it almost goes without saying that the very fact of companies being linked together across such a vast geographical area depended on the existence of the patent monopoly.

§

If patent-based relationships underpinned the corporate and financial structure of the telephone industry, they also had important implications for other "system" aspects of telephony—the physical, technical, and managerial ties between firms. Bell leadership was happy to credit the patent monopoly with lowering the costs of coordination, and thus smoothing the way to networks of greater scale. In the company's annual report for 1885, President William Forbes stated that "[f]or purposes of intercommunication between existing smaller systems, it is arranged that all shall work together, and the fact that the telephone companies throughout the land have been held under such a general plan by the force of a government patent, has been and is of the highest importance to the public, for in no other way could even the present condition of telephone development have been reached so soon, or without great confusion among competing companies."[53] The existence of basic cooperation, on the other hand, meant something very far from centralized management, or even unity of purpose among the industry's actors. "Working together" to coordinate telephony's industrial and technological development under the patents turned out to be a process of negotiation between parent company and affiliates.

During the industry's first stages, coordination was achieved by contract—that is to say, by stipulations built into the patent licenses. The number of restrictions and conditions placed on service by the Bell Company often came as a surprise to would-be licensees, many of whom had expected simply to "take the telephone, pay the rental, and use it."[54] Instead they found themselves bound to the strict leasing model of the patent-holding company, and, after 1880, subject to an increasing number of rules regarding interconnection with neighboring companies and toll line providers.[55] Contractual ties also governed American Bell's manufacturing arrangements. Although the company's close relationship with the electrical instrument supplier Charles Williams amounted to a degree of informal vertical integration, the Williams workshop quickly proved unable to meet the growing demand for equipment, and supply agreements were concluded in 1879 with four other manufacturers spread across the country. In theory, the contracts allowed Bell to control prices, quality, and terms of sale. In practice, such conditions proved difficult to enforce once the manufacturers began to compete with one another for business.[56]

At the center of these contractual relationships, American Bell itself did not direct telephone development in any great detail. Management in Boston was responsive to operating questions that arose in the field, but on many subjects had little (if any) more competence than its licensees.[57] The parent company maintained an Electrical Department run by Thomas Watson, which inspected manufacturers' output and labored on improvements to instruments and switchboards. The department's most significant work, however, consisted of evaluating outside patents for acquisition—a crucial function, but an essentially passive one.[58]

In 1882 American Bell altered this picture somewhat by recentralizing the manufacture of telephone equipment. The firm gained equity control of the Western Electric Manufacturing Company and licensed it as the exclusive supplier of telephones, switchboards, and other apparatus to the Bell operating companies. Chicago-based Western Electric was an important player in electrical engineering, having grown into the country's largest electrical manufacturer by supplying telegraph and telephone equipment to Western Union. To American Bell, Western Electric offered productive capacity high enough to solve the telephone company's perennial supply problems. It also represented a proven source of innovation and patenting that could become dangerous to the monopolist if not brought under control.[59] Bell's acquisition of the manufacturer in February 1882 created an important technical hub for the telephone industry. In the long term, the parent company formed a staggeringly successful symbiosis with its manufacturing subsidiary, especially in the development of a capacity for science-based research and development.[60] In the shorter term—that is, the patent monopoly period—American Bell gained a powerful partner in equipment innovation and a greatly enhanced capacity to ensure supply, monitor quality, and promote standardization.

The parent firm's growing vested interest in the operational side of telephony was apparent in its relationships with the operating companies. One issue during the mid-1880s particularly illustrates this point. During 1885, many of the telephone operating companies came under financial strain, a state of affairs that they ascribed in part to the high royalties imposed on them by American Bell. Bell leaders replied that the local firms had underestimated the costs of maintaining and updating their plant.[61] Nevertheless, in the face of price increases and subscriber losses, American Bell convened a conference of its licensees in Boston's ornate Hotel Vendôme, laid on "liquid refreshment and good fellowship," and renegotiated terms.[62] On a company-by-company basis, the parent firm made multiyear concessions in its royalties totaling about $200,000 per year—then around 10 percent of the total. Furthermore,

American Bell allowed operating companies to divert dividend payments on franchise stock to new construction, thus "sharing the cost of developing the business."[63] "The licensees should be led to feel," American Bell announced, "that it is for our interest as much as their own that they should be successful."[64] These measures represented not only a concession to temporarily adverse conditions, but also a decision to forgo immediate exploitation of the patents in favor of solidifying the telephone business.

The years that followed saw the rise of new commitments to integration on the part of industry executives, primarily in long-distance service and the standardization of equipment. In 1885 the parent firm established the American Telephone and Telegraph Company (eventually known as AT&T, although the abbreviation was not common usage at the time) as its "long-lines" subsidiary, charged with developing long-distance business. American Telephone and Telegraph operated a system of toll lines separate from the networks created by local operating companies. These routes employed new technology: "metallic circuit" lines of paired copper wires to replace the usual single-wire, earth-return method, which had proved inadequate for long-distance transmission. To use the service, existing telephone subscribers either had to rent a metallic circuit line for their own connections (at a rate considerably above that for local service) or else make calls from the long-distance company's own call office.

Despite this initial division of facilities, American Telephone and Telegraph became an engine of integration within the Bell group. Local companies, beginning with the big-city exchanges, came under increasing pressure to upgrade their lines and switchboards in order to interconnect with AT&T's metallic-circuit system.[65] Because intercity connections required local companies to coordinate charging arrangements and traffic management with the long-distance firm, AT&T became the center of Bell thinking about the organizational relationships between telephone companies. These features made American Telephone and Telegraph a natural corporate home for the advocates of "system" values. One of the prime movers behind the long-distance company was Theodore Vail, previously American Bell's general manager, who had been one of the first executives to envision the establishment of a long-distance network, and whose ambitions for regional interconnection ran substantially ahead of both the operating companies and the parent firm's relatively cautious leadership.[66] After Vail's resignation in 1887, other AT&T managers emerged to champion the long-lines company as an ideal instrument for gradually centralizing control of telephone operations. These men sought to take advantage of the patent monopoly—and prepare for the competition to follow—through an agenda of centralization, interconnection, and standardization.

This agenda had its limits. Despite the energy emanating from American Telephone and Telegraph, long-lines operations were marginal compared to the business done by the operating companies. In 1892, the year that AT&T opened its flagship New York-to-Chicago line, the long-distance company handled less than 1 percent of the estimated 600 million telephone calls made.[67] Operating companies dwarfed AT&T in revenue and capital investment. They also exhibited a considerable diversity of technical and managerial styles. Different types of market and levels of access to capital meant that telephone executives in Boston or New York perceived the ideal system very differently than did those in the South or the rural Midwest. The advanced switchboards, underground cables, and long-distance lines that increasingly represented the ambition of industry leaders in the urban Northeast had less appeal in areas where demand was less densely concentrated and where finance for new plant was scarce.[68]

American Bell and American Telephone and Telegraph perched on top of this diverse and decentralized group, with only limited ability to contribute to the shape that service took. To the extent that executives of the national firms could promote standards and practices among the Bell-affiliated group, they did so. But their efforts mainly took the form of presentations to annual industry conferences, principally the meetings of the National Telephone Exchange Association (NTEA) and two other forums known as the switchboard conferences and the cable conferences. Even there, American Bell's influence was limited: the NTEA, as the trade association of the local operating companies, treated the influence of Bell and Western Electric with wariness rather than deference.[69]

Where coordination worked best, it did so because of shared ideas about how the telephone industry should develop as a technical and a commercial proposition.

In 1894 the president of American Bell, John E. Hudson, claimed that instead of "a series of isolated exchanges or scattered lines . . . something very like a national exchange is showing itself."[70] This statement was a considerable exaggeration. Although many of the country's larger cities now possessed regional and interregional connections, integrated nationwide service would not be a reality until the twentieth century. Like Hudson, later observers would be tempted to project onto the telephone industry of the 1880s and early 1890s a cohesiveness that did not yet exist, and to seek in the system-minded programs of AT&T managers an emergent version of the later "Bell System." Shifting the focus away from the twentieth-century end-point, however, allows us to appreciate better the mixture of integration, cooperation, and autonomy that characterized telephony under the patent monopoly.

§

What emerges from the telephone story are the ways in which patents linked things together. Proprietary control spread out through technical and business organizations, spilling over from one technology and one firm to the next. A first set of spillovers occurred between the patented item and related products and services: in this case, between the telephone instrument and the systems in which it was embedded. In technical or economic terms, the role of the terminal telephone device in controlling service was a case of the tail wagging the dog. Telephones themselves were among the simplest components of the entire system, initially costing less to make than the call bells attached to them.[71] Switching and transmission were the major subjects of innovation, required the greatest investments by telephone providers, and ultimately determined the economics of service, yet the patents that controlled access to the telephone business were those for the transmitter and receiver. In part, this relationship between device and system was due to Bell's practice of leasing telephones: by selling a service rather than a stand-alone machine, the company packaged the whole hinterland of the network with the transaction. But it also reflected the power of patents over interdependent technologies. Cameras and film, music players and recordings, light bulbs and electrical generation all demonstrated the leverage that patent rights over part of an industry could exert over the rest of it. The competitive advantage gained by spillovers of patent rights to cover associated products was similar to (and often combined with) the effects of technical standardization, compatibility and interdependence, and even reputation and branding.[72] In these contexts, the influence of intellectual property rights is hard to distinguish from other pressures of technical integration and system management. Patents represented one strand in a tight web of technological control.

The basic telephone patents brought a proprietary dimension not only to wider technical systems but also to business organizations. Telephone companies, like other early electrical industries, "crystallized around patent rights," drawing their corporate organization in large part from the patent holders' strategies for exploiting their intellectual property.[73] This point is worth dwelling on. Prevalent economic theories of the firm posit that business organizations take shape by selecting from a range of coordination mechanisms: some functions are handled within the firm, through integrated managerial structures or "hierarchies"; some transactions are left to arm's-length market dealings; still others may be allotted to trade associations or other suprafirm bodies; and so the horizontal and vertical structures of an industry emerge. As the more nuanced versions of this model recognize, however, market/hierarchy decisions are not carried out in a vacuum, according to efficiency considerations alone. Instead they are made in specific tactical situations, with limited information, and influenced by factors ranging from social networks to cultural preferences.[74]

Once we broaden the focus from efficiency and transaction costs, we can see the kind of role that patents played. Patent rights, and especially patent monopoly, framed the choices of executives, conditioning their incentives, their opportunities, and their offensive and defensive strategies. Intellectual property also became a tool for managing transactions themselves—another mechanism of coordination lying in the spectrum between pure market dealings and integrated hierarchy. The "proprietary utility" model that arose in telephony and other electrical industries did so not because of those industries' natural tendency toward integration, but because of their persistent decentralization. Within this framework, the conditions attached to patent licensing and the corporate ties associated with franchise stock became important coordination mechanisms for the industry as a whole. The result was that patent monopoly left its mark on the companies that provided telephone service, even once the patent rights themselves ran out.

CHAPTER 8

Patents and the Networked Nation

THE BEGINNING OF THE 1890S saw the Bell telephone monopoly preparing for the end of its basic patents. Alexander Graham Bell's seventeen-year American grants would expire in March 1893 and January 1894. These dates were moments of truth for the telephone companies. Governments and corporations alike would now find out whether the monopoly model created by the patents was sustainable or appropriate for the industry once legal protections were gone.

On the eve of the Bell patents' expiration, America's telephone companies presided over roughly a quarter of a million telephones.[1] The record of growth to that date was mixed. Between 1880 and the end of 1884, the number of exchanges nationwide had risen from just over one hundred to more than nine hundred, and subscriber numbers had grown at an average of 25,000 per year. Since that time, however, expansion had slowed. An industrial depression, rising costs, and concern for their profits made the Bell-affiliated companies conservative in adding new customers. After 1885 the number of exchanges in operation actually fell, as facilities combined or closed, before rising far more slowly than before. Subscriber numbers edged upward at an average of 13,000 annually.[2]

On the other hand, user numbers were only one index of progress. Over the same period, the companies had amassed a harder-to-measure record of technical improvement. Early flat switchboards had given way to upright "multiple" boards that greatly increased the capacity and complexity of exchanges. Iron wires and wooden poles had been superseded in the cities by copper circuits and underground cables. American Bell was careful to note that telephone company investments went "not only to extending the reach of the service, but to improving in every way its quality and character."[3]

Both the moderate rate of diffusion and the keen focus on technical quality reflected the Bell companies' predominant business strategy. This consisted, in part, of a traditional monopolist's profit maximization by high prices and limited supply. Across the patent period as a whole, monopoly returns earned American Bell an annual average return on capital of 46 percent.[4] The parent company's dividend payments reached 10 percent of the face value of each share in 1883, and thereafter remained between 15 and 18 percent until the end of the patent period—by which point $25 million had been paid out.[5] Instrument royalties set in Boston helped to push the high-price strategy down through the operating companies.[6] Even after making payments to the parent firm, licensees generally enjoyed monopoly profits. Where they did not, as when companies expanded into small towns "probably . . . too rapidly" in the early 1880s, managers reacted by closing marginal exchanges and cutting back on construction.[7]

The Bell companies also adopted a particular set of assumptions—social and commercial—about the market for telephone service. Executives imagined a user base of businessmen, professionals, and service establishments, with a niche market for residential use by well-off households. They had good reason to think that demand for telephones would spread no further. Annual subscription rates ranged from $30 to $150 in most cities by 1893 and as high as $250 in the largest. At a time when the average nonfarm worker earned $450 per year, telephone use was beyond the reach of the masses.[8]

With the business market in mind, telephone companies adapted their service to a particular set of perceived consumer preferences. As American Bell's annual report put it, "the business man in this country, to whom the use of the telephone is so important a factor in the conduct of his affairs, wants the best and most extended service that means and skill can provide."[9] The companies' commitment to this type of service took concrete form in the development of their networks. Middle- and long-distance lines, for example, provided one of the most visible indicators of the companies' priorities. By the mid-1880s, all the Bell-affiliated firms and most of their local managers oversaw interconnection between neighboring towns and cities. The geography of these routes depended on local conditions. In the Northeast, great technical and organizational efforts were ploughed into building intercity lines several hundred miles long, meant to answer the demand of an elite customer base for communication with distant commercial centers. Managers in less urbanized parts of the country saw greater potential in shorter lines of fifteen to fifty miles. But even nonmetropolitan telephone men agreed on the importance of linking locations "intimately connected in business."[10] Programs of intercity building continued to gain momentum through the later 1880s and 1890s and eventually came to be regarded as emblematic of Bell's technological achievements and aspirations. American Telephone & Telegraph's line from New York to Chicago, which opened with great fanfare in 1892, was hailed as the "crowning achievement" of telephony under the patents.[11]

Glory aside, however, the most important network-building took place within cities—and especially underneath them. The later 1880s and early 1890s found the Bell companies heavily engaged in converting their largest networks from overhead iron wires to underground copper circuits. Underground construction was a specifically urban imperative. The proliferation of overhead electrical wires in major cities by the mid-1880s caused serious problems of induction on telephone lines, as well as public hostility to a "standing nuisance and source of danger." Telephone companies scrambled to develop cables that would work beneath the streets, hastening their efforts when cities began to pass laws forcing the wires underground.[12] Metallic circuits, meanwhile, embodied the Bell strategies of high quality and system integration. Not only did they provide superior service within cities, but they enabled the interconnection of local exchanges with AT&T's long-distance network. The growth of underground metallic-circuit lines became American Bell's chief

Alexander Graham Bell inaugurated the long-distance line from New York to Chicago in 1892. Bell himself had little or nothing to do with the telephone business after its first few years, but the Bell Company kept the inventor and his heroic act of invention at the forefront of its publicity. Library of Congress, Prints and Photographs Division, Gilbert H. Grosvenor Collection of Photographs of the Alexander Graham Bell Family, LC-G9-Z2-28608-B.

measure of the quality service that "the business or professional man" deserved.[13] As the telephone companies' largest class of plant expenditure, nothing better expressed the Bell commitment to "construction and equipment of the most costly nature."[14]

Telephone networks built on these principles were emphatically not designed to maximize the number of users. The Bell companies' investments in complex local systems and regional interconnection enabled them to penetrate urban markets and successfully establish themselves in business communications. But the effect of their strategy was to place geographic and social limits on the spread of telephony. In 1893, half of all subscribers resided in cities of more than 50,000 people, although these cities contained only one-fifth of the American population. Fully a third of all telephones rang within three hundred miles of Boston, in the most industrialized slice of the country. Across large tracts of rural and small-town America, Bell companies provided at best a scattered service; often, they explicitly refused small communities' petitions to build exchanges.[15] Even in the most-telephoned of the big cities, where hotels and newspapers might count dozens of telephones in a single building, the overall ratio of exchange subscribers to inhabitants was low: a generous estimate in 1895 put it at 1 to 150 in New York, 1 to 100 in Chicago, 1 to 70 in Boston. Nationally the figure was more like 1 in 250.[16] With the exception of well-off households and telephones placed in drugstores, the telephone had barely begun to diffuse into life outside the urban workplace.

§

Unmet demand and restrictive monopoly proved a volatile combination. The Bell companies' control over the price and availability of telephone service came under attack throughout the patent period.[17] As early as 1881, subscribers in New York City complained of high rates and poor service; subscribers in Washington, D.C., launched a twelve-day boycott, or "strike," to protest the local Bell company's switch from a flat-rate to a per-call charging structure. Disgruntled Philadelphia customers formed a protest committee the following year.[18] These telephone users believed that they were being gouged by repeated price rises and by rates much greater in larger cities than in smaller ones. In fact, both complaints had the same source: the increasing costs to scale that came with operating ever-larger and more complex switched networks. Because the number of possible connections increased geometrically with the number of users, each additional subscriber placed a greater demand on a telephone company's switchboards and operating staff. Big cities cost more to service than smaller ones, and rapidly growing exchanges were forced to raise prices repeatedly. Unfortunately for the Bell firms, they could never persuade their customers that costs should be higher, not lower, in bigger systems.[19]

By the time the telephone companies embarked on large-scale rate increases in 1885 and 1886, subscriber dissatisfaction was ready to explode. Users across the country embarked on a series of boycotts and rate strikes that shut down exchanges from Memphis, Tennessee, to Rochester, New York. Rochester's strike lasted eighteen months, from November 1886 until May 1888.[20] At least a dozen states and major cities proposed maximum-price regulations that the Bell companies considered punitive. One statewide rate cap passed: Indiana's $3-a-month rate law of 1885. The response of the two Bell-affiliated telephone companies covering Indiana was immediate and dramatic. With the approval of American Bell in Boston, the Central Union Telephone Company began to close exchanges—starting with Indianapolis, the center of the anti-Bell effort, where a quarter of the city's lines were disconnected within weeks. Cumberland Telephone and Telegraph withdrew from Indiana entirely, leaving the southern part of the state without service. For the four years the law remained on the books, torn-out telephone sets and dismantled exchanges in Indiana provided an object lesson in Bell's determination to control the field alone and on its own terms.[21]

Given the tide of frustration that the patents held back, it is hardly surprising that the threat of market entry pressed in on Bell at many points. Company president William Forbes told American Bell's shareholders in 1886 that "one decision by the United States Court invalidating our patent would be enough to flood the country with competing companies."[22] Basic telephones were easy to make and relatively cheap to set up. If users did not want or need the expensive and complex facilities in which the Bell companies had specialized, then small systems lay within the means of most communities. Entrepreneurs, mechanics, promoters, local subscriber committees, and roving purveyors of dubious telephone patents alike all saw a market in the shadow of the Bell patents.

It is impossible to know how many telephone providers operated without Bell permission during the patent period. Even the six hundred or so infringement suits initiated by American Bell cannot account for every backwoods exchange, jerry-rigged rural line, and workshop-built telephone that went unnoticed or shut down without legal action. On the other hand, a number of the six hundred defendants were companies that existed mainly on paper and never put a meaningful service into the field. It is clear that few non-Bell companies achieved any size or longevity as commercial enterprises.[23] The president of the National Telephone Exchange Association (NTEA) in 1884 declared it "perfectly safe to say that there is not in operation anywhere an exchange which is deriving a serious income from its business, or is affecting the income of any Bell Exchange Association seriously."[24] Most infringers preferred to work on the margins of markets occupied by Bell; they operated cheap, rudimentary facilities in the thinly served southern states, for example, in the upper Midwest, and in the anti-Bell hotbed of Indiana.[25] Pennsylvania was something of an exception: the state became a rare eastern center of infringement during 1883 and 1884, when its federal courts temporarily suspended Bell's litigation pending the resolution of the Drawbaugh case in New York.[26] Otherwise, companies that challenged Bell in its core urban and industrial districts could not hope to avoid injunctions. For the most part, enemies of the monopoly in these areas placed their hopes in patent pretenders such as Antonio Meucci (promoted by a committee of Philadelphia telephone users) or Sylvanus Cushman (backed by Chicago politicians and drugstore owners).[27] In the gaps between lawsuits, anti-Bell interests took every opportunity to circulate petitions, sign up potential subscribers, and demonstrate a constituency for service other than that of the hated monopoly.

In retrospect, it is easy to say that large-scale competition after the patents was inevitable. Yet the Bell companies remained sanguine as the expiry of their legal monopoly approached. Their commercial position had still not been tested directly. The incumbents believed, as did many observers in the electrical press, that Bell had the resources to withstand a siege. "There is a solidity about their telephone successes in the past," noted the *Electrical Review,* "that very properly gives them confidence in the future."[28]

Before this proposition could be tested, however, American Bell's leaders moved to protect their monopoly in the most familiar way: by prolonging patent control over the telephone. Their opportunity to do so rested on the two highly unusual delayed grants to Emile Berliner and

Thomas Edison. Both men had applied for patents on basic aspects of the telephone transmitter in 1877, only for their applications to remain in Patent Office limbo throughout years of trials and interference proceedings. In November 1891 and May 1892, the patents finally issued. If read broadly, each offered the possibility of locking competitors out of the standard form of transmitter. The temptation for American Bell to extend its fundamental patent control by up to fifteen more years proved irresistible. Bell's president at the time was John E. Hudson, who had been general counsel of the telephone firm as early as 1878 and who was by all accounts a cold and uncompromising representative of Brahmin Boston capital.[29] Hudson was a natural figure to continue American Bell's intransigent patent policy, yet his decision to reprise the firm's bitter campaign for legal monopoly cannot have been taken lightly. James J. Storrow spelled out the stakes: "The Bell Company has had a monopoly more profitable and more controlling—and more generally hated—than any ever given by any patent. The attempt to prolong it . . . by the Berliner patent will bring a great strain on that patent and a great pressure on the courts."[30]

Events proved Storrow right in the details and justified in his general pessimism. Pressure on Berliner's claim began even before the patent issued, when the *New York Times* called public attention "to this outrageous abuse of the privileges granted by the people to inventors."[31] Scarcely had American Bell declared its intention to enforce the Berliner patent than the company faced a government suit to cancel the grant. Under the government's charge of fraud, Berliner's rights were struck down in January 1895, reinstated on appeal in May, and finally cleared of wrongdoing by the Supreme Court in May 1897. Storrow himself did not live to see the Berliner patent upheld, having quietly passed away amid his law books in the Library of Congress just a month before the Supreme Court's judgment.[32] He would not witness American Bell's subsequent struggle to enforce the grant, which ended four years later when the U.S. courts in Massachusetts reduced the patent to its narrowest possible scope and thereby to irrelevance.

Bell's attempt to exploit the Edison patent followed a different track. Edison's specification covered the carbon transmitter, a subordinate invention to Berliner's microphone discovery but still a necessary component of the practical telephone. In essence, it was the same patent that had so effectively controlled British telephony for the United Company. Therein lay its great weakness, however. Any grant for an invention previously patented in a foreign country was deemed by American law to expire at the same time as the earliest foreign grant—meaning, in this case, at the completion of Edison's British term in 1891. Yet because Edison had *applied* for his American patent first and only *received* it after the foreign issue, legal authorities were unclear as to whether the foreign-expiry rule applied. To save the patent, American Bell had to fight on this point of law.

In doing so, Bell joined a contest that was already coming to a head. Systematic international patenting by American firms and inventors in the 1870s and 1880s, combined with lengthy examinations in the U.S. Patent Office, had placed several major patents in the same position. They included inventions in rubber manufacturing, electrical equipment, and, most notably, Thomas Edison's key electric lamp patent, now the property of the newly formed General Electric Company. With the involvement of the telephone and electrical combines, the *New York Times* estimated, "patents involving fully $600,000,000 of capital in their development and operation" were at stake.[33]

What followed was a two-front campaign to free those grants from the dead hand of their foreign counterparts. On the litigation side, American Bell and General Electric effectively took over a test case *(Bate Refrigerating Company v. Sulzberger)* then approaching the U.S. Supreme Court.[34] At the same time, they pressed Congress for a change in the law. The two companies made for a formidable lobby and succeeded in placing their proposed amendment in five House bills between December 1893 and April 1894.[35] Neither the Court nor Congress proved obliging, however. In early 1895, the justices issued their decision declining to read the existing statute as the telephone company wished.[36] The legislature meanwhile failed to pass a patent law amendment in timely fashion, repealing the foreign expiry provision only as part of a broader bill in 1897. Reform came too late for the Edison telephone patent, which had to be given up for dead when the Supreme Court handed down its *Bate* ruling.[37]

The Bell Company did still have other patents at its disposal—approximately nine hundred of them, in fact.[38] Since 1879, the company had maintained a Patent Department, led by the engineer and attorney Thomas Lockwood, whose remit was to secure patents on internally generated technologies and to scour the market for telephone-related inventions by others. Lockwood closely scrutinized the electrical patents emanating from the Patent Office and examined "patents or inventions submitted by the public for consideration." Acquiring such patents was made considerably easier by the fact that the Bell-affiliated companies were the only plausible large customers for telephone improvements during the monopoly period.[39] Internal invention was less favored. Lockwood was initially convinced "that it has never . . . and never will pay commercially, to keep an establishment of professional inventors" on staff. He did, however, make a practice of filing patent applications on all Bell employee inventions, with little regard to "merit, and the presence, or extent, of invention."[40] To the extent that there was a real inventive powerhouse within the Bell group, it was the manufacturer Western Electric, and above all its engineer Charles E. Scribner. Scribner amassed more than 400 patents during his career, many of them on the multiple switchboard technology that was the acme of technical accomplishment in the early telephone industry.[41]

Bell drew on the remainder of its patent portfolio to try to ward off competition, but only with limited success. Western Electric won suits against non-Bell exchanges under a variety of switchboard patents.[42] Other suits were "vigorously prosecuted" over what a later government investigation called "telephonic appliances."[43] At best, though, these actions could impede rather than shut down competing telephone companies. Worse, the Bell Company felt the tide turn with its first significant loss in a patent infringement suit, when the federal court in Chicago struck down Thomas Watson's "switch hook" patent in 1897.[44] The Bell side's greatest hopes had been placed in the Berliner and Edison gambits, and these proved to be ill-fated rearguard actions. They say much about the grim determination of American Bell to sustain its monopoly and to cleave to the patent-centered strategy that had served the firm before. But as barriers to entry, both patents foundered. "In a word," the unsympathetic *Western Electrician* noted, "the American Bell Telephone company is not feared as it was in the early days of the art."[45]

The sight of Bell on the legal defensive did little to discourage rival companies from planning their entrance. Quite the reverse: the campaign against the Berliner patent galvanized the pioneers of an independent telephone movement into organization and public agitation. The Telephone Protective Association, founded around 1895 as a patent defense fund for non-Bell equipment suppliers, would go on to become the National Independent Telephone Association, one of the principal trade groups for Bell's new competitors.[46] Independent telephone companies began to trickle into the market despite the looming Berliner and Edison patents: Census Bureau figures, certainly undercounting, show twelve new companies in 1892, eighteen in 1893, eighty in 1894.[47] The Supreme Court's decision dooming Edison's transmitter patent in March 1895 signaled the beginning of the rush. "It is on the strength of this decision," stated a Baltimore telephone manufacturer, "that all the independent telephone companies have been started."[48] Two hundred companies appeared in both that year and the next, followed by progressively larger numbers thereafter. American Bell's legal control over telephony was at an end.

§

The 1890s and 1900s demonstrated both how important the fundamental patents had been to the Bell monopoly and how deeply the monopoly had shaped telephone service. A decade after Alexander Graham Bell's rights ran out, the telephone business was open to all comers and had changed spectacularly as a result. Thousands of new telephone providers and millions of new users showed just what expansionary potential the iron grip of the patents had previously held in check. Meanwhile, the raging competitive battle between Bell and the independents provided a test—the audit of war, to borrow a phrase—of what exactly the patent years had done for the incumbent companies.[49]

In 1894 plenty of informed commentators believed that the patents had given the Bell firms a lasting stranglehold on the telephone market. Against established networks in and between cities, even Bell's old adversary Grosvenor Lowrey "doubted if a new company could accomplish much by taking the field." As Lowrey explained, "The value of a telephone in my house is based on the subscription list of the company. The instrument of a company with 1,000 subscribers is twice as valuable to me as that of a company with only 500 subscribers."[50] This view fit perfectly the conservative mindset of American Bell's leadership. It required only that the telephone companies trade on their existing lead in facilities and technology, and continue to expand their network of intertown connections—in other words, to carry on doing what they already did well.[51] The implicit assumption was that new entrants would compete on Bell's terms.

Received wisdom in Boston turned out to be spectacularly wrong. New telephone providers flourished by reaching markets other than the ones Bell occupied, adopting alternative technical styles and providing a product that sometimes barely resembled telephone service as Bell knew it. Between 1894 and 1902, at least nine thousand of these independents began operations, more than half of them "farmer lines" without exchanges of their own.[52] Of the remainder, around three thousand were "commercial" systems run for profit; one thousand more were "mutual," or cooperative, enterprises owned by their users. Many of these undertakings had little in common with one another; they ranged from isolated rural lines to substantial urban exchanges. But the new entrants collectively departed from the Bell model in several crucial respects.

In particular, the independents made their fastest progress in areas that the Bell firms had covered relatively lightly: rural regions, smaller towns, and states outside the industrialized Northeast and Mid-Atlantic. The comparatively poor South, for example, gave rise to many more independent ventures than did Bell's strongholds in the Northeast. The Midwest produced most of all. Populous and studded with small towns, with a mixture of industrial cities and an increasingly mechanized rural economy, the region offered promising markets for new telephone enterprise.[53] When further nourished by political hostility toward Bell's highhandedness, the Midwest proved to be "fertile soil for anti-monopoly sentiment."[54] Agricultural Iowa had the largest number of independent providers in 1902 (more than fifteen hundred), followed by Indiana, Illinois, and Missouri.[55] Almost three-quarters of the new entrants germinated in twelve midwestern states, many in communities hitherto barren of service.

The independents also offered a new approach to service: what one of their early publicists called a "bottom-up" rather than a "top-down" style.[56] The most salient features of the new companies were simpler technology and lower prices, which differentiated them from their Bell counterparts and enabled them to thrive in previously marginal markets. In addition, though, the independents' signature skill lay in creating new universes of connections. Whether they were rural mutuals or urban commercial enterprises, the new providers connected places that Bell had never joined together: farms to nearby towns, cities to their suburbs.[57] Independent telephone entrepreneurs made much of their connection to local communities; they called their businesses "Home" companies and damned Bell as a creation of Eastern money interests.[58] But they backed up the rhetoric with denser networks, tributary connections, and assiduous interconnection with tiny rural systems. By making a virtue of localization, the independents were able to satisfy a clientele that was basically localized in its interests.

Moreover, the independents could and did challenge Bell companies on their own ground. Although failure to reach a broader market had given the Bell monopoly its soft underbelly, market segmentation was not the only factor creating space for an independent challenge. Companies entering the market against Bell were also able to exploit structural features of telephony that favored competition. The greatest of these was the diseconomy of scale in exchange operations, which placed Bell incumbents at a cost disadvantage against their smaller rivals. Although Bell companies lowered prices after the patents ran out—average revenue per telephone fell from $90 per year in 1893 to $63 in 1900—they did not generally match the low rates offered by the independents.[59] Once competition began, Bell managers from one middling town after another reported that "[w]e cannot meet these [independent] rates, and cannot sell our metallic service at the present rate . . . in exchanges of that size."[60]

Technical change mattered too. In the first quarter century of commercial telephony, plant turned over so fast that sunk costs quickly became burdensome legacies. Managers in regional cities such as Milwaukee and Buffalo set the life expectancy of telephone plant at five years; many of them completely rebuilt their exchanges three times during the 1880s.[61] The *Electrical Review* claimed in 1892 that "not a foot of wire or a section of switchboard which was in use six years ago is now in service in New York."[62] New providers entered the market unencumbered by legacy costs of capital: by 1902 independent companies had invested an estimated $192 per telephone, compared to $328 per telephone invested by the Bell firms.[63] Meanwhile the Bell companies responded to creative destruction by providing large amounts for depreciation—up to one-third of gross receipts in some cases—further widening the price gap between Bell and independent service.[64]

By the turn of the century, direct competition was in full swing. Of the nine hundred or so American cities with more than five thousand inhabitants, almost a quarter had competing telephone systems in 1897; more than half did so in 1902.[65] In this era of "dual service" it was not uncommon for telephone subscribers to maintain two phones—one connected to the local Bell exchange and one to its independent rival.[66] The distinctive nature of network competition gave the period a special place in the growth of telephony. As companies vied with one another to offer the most comprehensive universe of connections, they spurred what historians have called "access competition" or "cutthroat competition": a frantic and unstable race for network superiority.[67] Bell employees with long memories might have known what to expect. During the struggle with Western Union in 1878–1879, one Bell figure recalled, "The cry was—cover the ground; get your lines up—give us good construction if possible, but give us the lines good or bad. . . . Rates were given, upon which the lines certainly had to be operated at a loss."[68] Bell companies would relearn all these methods after 1894. They learned, too, the truth of another such recollection: "[A]lthough such competition was severe for those engaged in it, it became a means for spreading the use of the telephone widely across the land . . . which no other agency could probably have equaled."[69]

Sure enough, competition after the patent period produced unprecedented levels of growth. Between 1894 and 1899, the number of telephones in America more than tripled, to over one million. Five years later the total was over three million; another half decade put it at seven million.

Both commercially and socially, American telephony had grown beyond recognition. Telephone companies *employed* more than 130,000 people in 1907—an army of operators, linemen, managers, and engineers ten times larger than the workforce of a decade earlier.[70] These men and women operated a service that was now ubiquitous in many parts of the country. By 1907 there was one telephone for every fourteen people in the United States.[71] In states and cities where direct competition had taken hold, numbers were still higher. The midwestern cities of Indianapolis and Cleveland had respectively one telephone for every ten residents and one for every eight; Iowa and California boasted one telephone for every seven inhabitants.[72]

The transformation of telephone service was qualitative as much as it was quantitative. Freed to compete and market aggressively, companies on both sides rolled out new products: cut-rate "kitchen telephones" for placing orders with local tradesmen; "party lines" by which two, four, even ten households shared a single circuit for a much reduced subscription; "nickel-in-the-slot" telephones, which replaced monthly rates with a single coin-drop per call. Installed in massive numbers in private residences and apartment buildings around the turn of the century, this last type placed telephone service within the reach of anyone with five cents to spend.[73] The nickel telephone and other devices were marketing developments more than they were technical ones. Inside its casing, the individual telephone of the 1900s contained the same basic technology as the device of the 1880s and early 1890s. But the sheer diversity of its settings and uses distinguished the early twentieth-century telephone from its antecedents. The elite instrument of 1876 to 1894—what the historian Robert MacDougall has called the "Boston phone," after the headquarters of its patent-wielding masters—was already a fading memory.[74]

Instead of ratifying the monopoly structure that prevailed under the patents, the years after 1894 revealed the weakness of nonlegal barriers to entry in telephony. Competition proved not just possible but vigorous, commercially viable, and politically popular in many parts of the country. Yet the rise of the independents did not sweep Bell away. In much of America, and particularly in some of the key urban and business markets, Bell firms remained dominant. Furthermore, the Bell group succeeded in rolling back the early gains of the independents. The proportion of non-Bell telephones nationwide peaked at just over 50 percent in 1908, after which the Bell companies began to recover market share. By the 1920s the former monopolist had regained something like its former supremacy: large numbers of independent systems had been bought out, driven out of business, or refashioned into appendages of Bell's network.

Bell's victory over the independents presents a complex saga.[75] Competition was waged in thousands of American communities from farm towns to great cities; it was fought in the press, in state legislatures and municipal governments, in the palatial offices of great capitalists like J. P. Morgan (whose syndicate took control of the parent Bell company in 1907), and in shabby hotel rooms where Bell agents struck secret deals with their supposed competitors. Throughout this process, the survival and recovery of the Bell confederation as the dominant force in the telephone industry could not be guaranteed. Nevertheless, path dependence played some role in the continuing dominance of the former monopolist. Market power established under the patents turned out to have a number of persistent effects.

The fruits of monopoly were clearest where Bell companies held their ground against challengers. Bell operators continued to provide a large majority of total telephones in the New England and Middle Atlantic regions. They also continued to enjoy an advantage in major cities. These locations had attracted the greatest telephone development during the patent period, and the erstwhile monopolists now reaped benefits from their large head start in intra- and intercity connections. In addition, Bell companies had entrenched themselves both literally and figuratively in strategic urban areas. In New York City, for example, the telephone company owned the Empire Subway Company, which controlled the city's underground conduits.[76] In this instance and others, the monopolist strenuously opposed the grant of municipal franchises to new telephone companies. A combination of commercial incumbency and political maneuvering made independent entry into the most valuable strategic markets extremely difficult. No independent operator ever succeeded in entering New York City, the "keystone" of the Bell System.

Chicago, the independents' Holy Grail in the Midwest, experienced only a token non-Bell service.[77]

Beyond these strongholds, monopoly power had national as well as local application. The Bell group was still a loosely coordinated assemblage of firms when competition began. Even so, the organization built under the patents proved capable of cushioning the impact of competition and of channeling resources to the most threatened companies. Cross-subsidy between firms achieved particular importance in the Midwest, where the heaviest competition took place. The Central Union Telephone Company, Bell's beleaguered affiliate for most of Ohio, Indiana, and downstate Illinois, survived only thanks to injections of capital from the parent firm. Fighting on the front lines of competition in the Midwest, Central Union lost money every year between 1896 and 1913, while the Boston firm poured in financing to the tune of $30 million.[78] Opponents believed that Bell's policy amounted to predatory pricing—an attempt, in the words of one Central Union director, "to cause every dollar invested in Independent property to be lost."[79] Thanks to the group's national financial and organizational resources, Bell was able to replicate the pattern elsewhere: in the South, for example, where the parent company underwrote aggressive pricing and new construction; and in upstate New York, where operating companies worked at a loss in order to deny the independents an approach to New York City.[80]

Financially, this counteroffensive transformed the Bell group. The loose corporate ties that joined the various companies together suddenly snapped taut: the parent firm owned every share of Central Union stock by 1913; across the operating companies overall, its holdings of licensee stock climbed from 45 percent of the total in 1900 to over 80 percent in 1910.[81] Massive capital requirements soon affected the parent company itself. New investment strained at the limits of American Bell's authorized capitalization before the 1890s were out, leading the firm to reorganize in New York in 1900 under the name of the American Telephone and Telegraph Company (AT&T).

Thus armed, much of Bell's strategy continued to be about using size to overpower the independents. Connection to large networks was not nearly as prohibitive an advantage as American Bell's leaders had expected, but it remained an element of competition. AT&T and the operating companies invested heavily in the development of intertown toll lines, thus increasing the value of local Bell service. The independents tried to respond by forming regional associations and long-distance companies, but Bell firms retained a significant advantage in coordinating regional networks.[82] In many ways, the independents gave a glimpse of what telephone system-building might have been like without the patents: a process of constant negotiation and accommodation among large and small operators.[83] Their method was viable, but complicated: as one weary independent manager put it, "The trouble with farmer companies is that they get what is commonly termed 'the big head' . . . they think they can dictate any terms and there shall be no deviation from those terms whatever."[84]

Bell's advantages of financial and organizational scale reached their greatest importance when the Bell firms began to purchase or co-opt key independents. After 1900, and more enthusiastically after 1907, AT&T allowed operating companies to sublicense noncompeting independents, attaching them to Bell networks and effectively making them into client companies.[85] By ceding marginal markets to sublicensees, Bell firms consolidated their hold on more lucrative territories, while simultaneously disrupting independent plans to form a coherent competitive front. Despite impassioned pleas for solidarity, independent alliances proved highly vulnerable to defection. One low point came when the president of the National Independent Telephone Association sold his Indiana firm to Central Union in 1905, leading the independent trade journals to howl at his "nasty treachery" and brand him "a traitor in the meanest form."[86] Within a decade, however, the number of voices holding out against cooperation with Bell had dwindled to an angry few. With assistance from a wave of financial failures among the independents during the first decade of the 1900s, Bell companies brought more and more systems into the fold through acquisition and interconnection agreements.[87]

Through these policies, the Bell interests began to return telephony to a noncompetitive, single-service model. The breadth of their strategy was exemplified by its accompanying political campaign. In the cities, Bell companies drew support from the natural constituency for single service: the business users whose desire for a broad universe of connections tended to be greatest, and who in most cases had borne the burden of maintaining two telephones under dual service. As an influential report of the New York Merchants' Association put it, "Competition in telephone service does not offer a choice of benefits, but compels a choice of evils—either a half service or a double price."[88] Such sentiments enabled Bell incumbents to oppose independent franchises in individual cities and gradually to roll out the argument for single service in regulatory forums across the country. In particular, the telephone companies cultivated the new state utility commissions that began to appear in the 1900s and 1910s, and which were less sympathetic to locally rooted independents than were municipal governments. Bell firms were able to persuade state regulators that the industry was a "natural monopoly," appropriately governed by compulsory interconnection and stringent quality standards.[89] As a result, regulation became another force promoting Bell's vision of service and ultimately cementing Bell's domination of the industry.[90]

§

No single theory explains whether and how broad patent monopolies generate lasting market power. Instead, as economists and historians alike have pointed out, the enduring importance of the temporary monopoly depends on the use that companies make of it. According to Alfred Chandler Jr., nineteenth-century pioneers of high-technology products cemented their dominance by building research organizations and by exploiting their economies of learning.[91] Other research has pointed to underlying "technological competencies" as the basis of a patent monopoly's prospects after protection runs out.[92] Sources of first-mover advantage can readily be slotted in according to the character of the industry or the theoretical predisposition of the author: the New Dealer Walton Hamilton, for example, blamed the oppressive "techniques of high finance" for the Bell System's long-term entrenchment after the patents expired.[93] Likewise, failure to sustain monopoly seems to have had many fathers. Where patent holders' market share has collapsed on the expiration of their rights, historians have identified proximate causes ranging from inadequate product development (Xerox in the 1970s) to the distraction and exhaustion that pioneers suffered in prosecuting their patents through the courts (the Wright brothers' aircraft business).[94]

Opportunities for prolonged control, then, rested on the patent monopolist's willingness and ability to exploit first-mover advantages. By implication, broad patents had greater economic implications where an industry was subject to path dependence or high barriers against later entry. Network and system industries certainly fit this profile.[95] Where manufacturing was concerned, Chandler may have been right that "power rested far more on the development of organizational capabilities than on creating 'artificial' barriers to the allocative effectiveness of market mechanisms such as patents."[96] But the nature of market dominance was inherently more "artificial" in network industries, where the efficiency imperatives and product substitution that disciplined manufacturing oligopolies were overshadowed by legal and political barriers to entry, technological path dependence, and sunk costs.

In early telephone service, the economics of first-mover advantage proved to be mixed. Certain users preferred the larger networks that incumbents could provide. But in other instances barriers to entry were not particularly strong, either because users preferred low prices to large, high-cost networks or because underserved markets offered a foothold for new entrants. For this reason, notwithstanding the technological competencies that the Bell incumbents built up during the patent period, it was a different set of factors—political, financial, and ideological—that determined their fate after legal protection ran out.

Conclusion

On November 23, 1936, Washington, D.C., hosted another patent centennial, this one marking the hundredth year of the U.S. Patent Office. Like the 1891 gala in honor of the first patent law, this event attracted cabinet officials, inventors, and lawyers from around the country. Once again, speakers lauded the leading inventions of the day and celebrated a patent system that had "served as a model for the world and made possible unified, coordinated progress toward happier living for all peoples."[1] Novelties on display included a rayon-clad "Maid of Science" modeling a synthetic silk purse made from the gelatine of sows' ears, as well as "Rubinoff, the noted radio artist," serenading attendees on his collapsible, pocket-sized, patented violin. For the day's finale, guests at the evening banquet—joined by the radio audience of the National Broadcasting Company's Red Network—were treated to the most elaborate performance of all. From a Douglas airliner circling in the skies above Washington, picked out by the "slender white beam" of a spotlight from below, the "Voice of Progress" intoned a list of the nation's twelve greatest inventors. Over the rolling drums of an army band, listeners heard the first name loud and clear: "Alexander Graham Bell, who gave us the telephone."[2]

Bell's identification with world-changing invention in general and with the American patent system in particular continues to this day. If nothing else, the inventor's legal victories secured his place in historical memory. Had Bell lost his claim in the U.S. courts, American schoolchildren might now study Elisha Gray or even Daniel Drawbaugh—though probably not the German Philipp Reis—in his place. To some extent, nations tend to claim their own: German authors have been quick to recognize Reis, while Italian and Italian-American groups have kept Antonio Meucci's candidacy alive.[3] In 2002, a largely Italian-American effort led the U.S. Congress to acknowledge Meucci for his "work in the invention of the telephone." Canada, where Bell lived for much of his adult life, immediately reaffirmed its adopted son's priority by Parliamentary motion.[4]

Popular histories are not the only ones that might have been written differently. Bell's life and thought have featured prominently in scholarly treatments of invention, down to a very fine level of attention to his personal circumstances.[5] The authors of these works probably do not consider themselves the captives of lawyers from a century past. Yet their accounts rely heavily on the huge mass of detail gathered for the legal record; on Bell's court-endorsed status as the genius behind a radical inventive step; and ultimately on the inventor's fame, perpetuated by the corporate empire built on his patents.

This last link is important. Patents were ultimately a tool of business, not simply a means to gain inventive credit, and they exist as landmarks in the history of technology only because economically motivated actors were willing and able to seek particular articles of intellectual property at particular times. Whether in search of heroic figures or aggregate statistics, historians must constantly remind themselves that the patent record is not the record of inventions, only of attempts at appropriation. The telephone in the United States and Britain provides a case in point. Whoever conceived the invention, it was the pioneering companies and their lawyers who truly "made" the fundamental patents: they who secured legal victories against rival claimants; they who shaped the scope of the Bell and Edison grants during litigation—including by emphasizing features that the inventors themselves had not believed to be central. In return, the patents underpinned the companies' commercial success, with predictable consequences for the historical reputations of the patentees. It was no coincidence that in 1936, when the organizers of the Patent Office centenary chose Alexander Graham Bell as the first great inventor on their roll of honor, the Bell System was the largest business organization in the world.[6]

Alexander Graham Bell did not live to receive the accolades of the 1936 centennial; he had died in 1922. Like every other inventor personally lauded at the ceremony (apart from Orville Wright, of airplane fame, who lived until 1948), Bell was a figure of the past. New inventions had their due at the 1936 event but, unlike in 1891, living scientists did not receive lengthy tributes from the stage, and none of the cutting-edge devices on display bore the name of an individual creator. Part of the reason could be seen in the promotional films that screened throughout the day. Their makers and principal subjects were the real stars of the event: Du Pont, General Electric, AT&T, Chevrolet Motor Co., and a handful of other companies and trade associations.[7] Corporations, not individuals, now stood at the forefront of invention and patenting.

Few developments in the social organization of technology have attracted more notice, or more concern, than the corporate control of invention. During the twentieth century, the meeting of big business and scientific technology became a central element of modernity and affluence, delivering a flow of new goods and services that transformed industrialized societies.[8] At the same time, however, corporations accumulated immense economic power by displacing craftsmen and independent inventors from the vanguard of innovation. The American economist Walton Hamilton, writing in 1941, memorably described the transformation under the heading "Technology swaps masters."[9] Hamilton's contemporaries gave voice to the anxieties that accompanied its progress. "[I]n an era that increasingly lives by science and technology," wrote one, the sociologist Robert Lynd, "business control over science and its application to human needs gives to private business effective control over all the institutions of democracy."[10] Even at the 1936 centenary—otherwise an uncritical celebration of industrial progress—speakers acknowledged the specter of corporate monopoly and its discontents.[11]

These mid-twentieth-century observers identified patents as a key instrument of corporate technological control. Defenders of the patent system pointed to its function in the encouragement of research, whereas critics focused on the anticompetitive uses of the grant, arguing that corporations had subverted the system of rewards to invention. In many ways, this difference in viewpoints represented an old debate within the Anglo-American patent tradition. The balance between legitimate property rights and excessive monopoly power, between the incentives provided for inventors and the opportunities created for rent seeking, had been under scrutiny since the eighteenth century. Yet those who saw the patent system transformed by big business had a point. The visible hand of managerial capitalism, with its ability to organize both innovation and market structure, had altered the power and the role of patents.

The story of the telephone patents was a notable episode in the changing relationship between patents and industrial power. It straddles the historical watershed between the world of independent inventors, who reaped such temporary monopolies as they could from their own patents, and that of their corporate successors: the innovation-oriented firms that sought sustained market control through a mixture of continuous research, patent protection, and integrated systems management. American telephony seemed to present ideal types on each side: on the one hand, Alexander Graham Bell, a lone inventor in the heroic nineteenth-century mold; on the other, American Telephone and Telegraph, the heir to Bell's original company and a pioneer of twentieth-century industrial research.[12] These ideal types miss a big part of the story, however. In between Bell's act of invention and the establishment of AT&T's industrial laboratories, the early telephone companies combined an aggressive legal strategy with proprietary corporate organization to produce the outstanding patent monopoly (legal or economic) of the second industrial revolution. Corporate power and intellectual property were thoroughly entwined before the rise of industrial research.

Patents thus help to explain how the second industrial revolution of the late nineteenth century could be both a "golden era of . . . independent inventors" and a phenomenon defined by organizational empire-building.[13] Alexander Graham Bell and Thomas Edison belonged to—indeed, were emblematic of—the golden-age generation. Like contemporaries such as Guglielmo Marconi and Nikola Tesla, they produced radical new technologies outside the direction of existing firms. When it came to commercializing their discoveries, patent property became central to increasingly ambitious plans to develop and dominate national markets. At first, patent-holding companies experienced a form of "personal capitalism" tied to ownership of intellectual property, with the role of inventors and their close associates strong enough to affect governance. Quickly though, the needs of capital began to take precedence in the form of patent-pooling, mergers, and the adoption of holding-company-like organizations. The result was an initially decentralized business model built around patents and defined by proprietary links. These interlinked firms formed the basis of the industrial giants to come.

In the case of the telephone, basic patents continued to play a formative role as the early telephone start-ups grew into much larger companies. A legal monopoly that lasted into the 1890s obviously operated to set the terms of telephone provision and had a powerful restraining effect on the growth of the industry. The influence of broad patent rights also went well beyond their effect on competition: licensing and the exchange of franchise stock provided a subsidy to the financial integration of the industry, as well as a mechanism for coordinating operations and technical standards. The networks and systems that defined telephony came accompanied by a whole variety of arguments for centralization and the suppression of competition among telephone companies. But the success of these arguments and the telephone pioneers' interest in making them depended heavily on the preexisting monopolies created under the patents.

The formative industrial role of the telephone patents depended, of course, on the success of legal strategies for asserting broad control under a pioneer patent. In legal terms, the telephone patent monopolies were not exactly "modern." Broad claims over radical new technologies were a phenomenon that Charles Goodyear, Thomas Blanchard, or even James Watt would have recognized. Furthermore, the telephone suits raised well-worn questions of patent law: priority of invention, adequacy of the specification, and the scope of property rights allowable to a radical new departure. Corporate change manifested itself in these cases not by raising new legal issues but by reshaping the old ones, partly through the growing complexity of scientific evidence, but mostly through the strategic power that large companies could bring to their litigation.

Above all, the telephone cases show the pressure exerted by market power (and its attendant politics) on a body of law that did not formally claim to regulate market outcomes. Patent law focused—at least in theory—only on matching the scope of property rights to the invention specified. Yet when the telephone patents came before the courts, technical arguments about priority and scope became infused with loaded issues of monopoly and antimonopoly. Under American law's search for the first and true inventor, the plain mechanic Daniel Drawbaugh came to represent hopes for a more popular and unrefined telephone service than the unbending Boston corporation would allow. Amid the more pedantic twists and turns of British patent adjudication, Edison's diaphragm bore the weight of hostility to the monopoly. The American practice of government suits for supposed fraud, which came to its greatest height in attacks on Bell and Berliner, exemplified the ways in which monopoly challenged the patent system and received both a legal and a political response.

This story was a product of its times. Like many other areas of nineteenth-century law, from corporation law to regulation and antitrust, patent law was a field in which the rising scale of business organizations attracted both legal support and legal resistance. In the telephone cases, powerful firms benefited from the tendency of the courts to support broad fundamental patents, especially in the case of patentees who could be presented as "great inventors." Numerous technologies of the second industrial revolution saw similar attempts to seize control of the field. Even as those attempts were under way, though, times were changing. Pioneer patents, always rare, had already begun to cede the spotlight to other instruments of control by the closing years of the century. Powerful though a patent for fundamental invention could be, reliance on a stand-alone piece of intellectual property was inherently risky, especially when the political temperature began to rise. By contrast, an arrangement of interrelated patents offered greater legal security, could expand to cover subsequent improvements, and could allow proprietary control of a technology without any of the grants involved possessing exceptional breadth.

Late nineteenth-century patent-using corporations—including many of those which had themselves been built on pioneer patents—thus tended to expand their range of intellectual property strategies over time.[14] Organized industrial research supplemented and partly supplanted the older practice of buying up promising patents from outside the firm. Patent pooling also took off. Firms had created large-scale cooperative patent structures as early as the 1850s, with the "Albany agreement" among American sewing-machine manufacturers.[15] The end of the century saw this approach widely applied in the increasingly oligopolistic electrical sector, most notably in the 1896 creation of the Board of Patent Control between General Electric and Westinghouse.[16] At the same time, the use of tying agreements and other patent-based restraints of trade expanded with largely favorable treatment from the courts.

Under such methods, patent monopoly during the twentieth century came to depend less and less on the content of any individual patent and more on the momentum of existing financial and corporate power. Successful patent exploitation had always favored the well-resourced—those with access to the courts and to the best lawyers, and possessed of the ability to bleed lesser opponents dry with lawsuits—but the rise of corporate oligopolies in the later stages of the second industrial revolution ultimately reduced the prevalence of patent litigation. As bureaucratically managed technology replaced the nineteenth-century free-for-all among companies, independent inventors, and speculators, the uncertainty and opportunism that drove much patent litigation diminished. Corporations turned to more predictable means of technological competition. AT&T, despite holding basic patents over inventions such as the loading coil and vacuum tube, brought almost no patent suits after 1908.[17] When the next great phase of electrical patent conflicts arose over radio in the 1920s, the dominant companies in the field chose to partition electrical technology rather than compete over the "no-man's land" between their patent portfolios. A series of grand cross-licensing agreements left each secure in its own fiefdom: General Electric and Westinghouse in power and light; RCA in radio broadcasting; and AT&T in wire communications.[18] The ability to accumulate patents and to incorporate them into a broader scheme of technology management had displaced the judicial definition of rights as the key to market control.

In the early twenty-first century, patent litigation has made a comeback.[19] This development partly reflects a reversal of the early twentieth-century concentration of inventive activity. Many high-technology companies have outsourced their research functions to specialized, vertically disintegrated firms. Meanwhile the independent inventor—never entirely absent, even in the age of corporate research and development—has reemerged as a character on the technological stage, now often backed by venture capitalists or by companies that specialize in acquiring and enforcing patents.[20] The litigious turn stems from several developments that echo the late nineteenth century: a rapid growth in the number of patents issued (with attendant criticism of lax examination standards); extensive changes in patent policy and the organization of the courts; and even attempts to assert fundamental patents over the key technologies of a "third industrial revolution," based on information technology and biotech.[21] Amid the new ferment of legal contests, critical observers have discovered a growing "dark side of patents."[22] Overbroad grants and patent thickets have flourished under strong enforcement. Worse still, the winners "are sometimes those with the best lawyers, or those simply lucky enough to have been awarded a key patent they did not really deserve, rather than those that have created the best products or services."[23] Contemporaries of Alexander Graham Bell might well sympathize with these complaints. Then as now, the widespread pursuit of economic power through the courts placed the patent system itself on trial.

Notes

Acknowledgments

Index

Notes

Abbreviations

F.	Federal Reporter
F. Cas.	Federal Cases
L. R.	Law Reports
L. R. App. Cas.	Law Reports, Appeal Cases
L. R. Ch. D.	Law Reports, Chancery Division
L. T.	Law Times Reports
NBER	National Bureau of Economic Research
PP	U.K. Parliamentary Papers
R. P. C.	Reports of Patent, Design and Trade Mark Cases
W. P. C.	Webster's Patent Cases

Archive Sources

AT&T Archives	American Telephone & Telegraph Company Corporate Archives, Warren, New Jersey
Bell Papers	Alexander Graham Bell Family Papers, Manuscript Division, Library of Congress, Washington, DC
BT Archives	British Telecom Archives, Holborn, London
Edison Papers	Thomas Alva Edison Papers, Rutgers University, Piscataway, New Jersey, digital archive at http://edison.rutgers.edu
Joseph Bradley Papers	Joseph Bradley Papers, New Jersey Historical Society, Newark
NARA	U.S. National Archives and Records Administration, College Park, Maryland

NARA Philadelphia   U.S. National Archives and Records Administration, Philadelphia, Pennsylvania
UKNA   U.K. National Archives, Kew, London

Introduction

1. The following description is drawn from "The Patent Centennial Celebration," *Scientific American,* April 18, 1891, 243–245; "One Hundred Years," *Washington Evening Star,* April 8, 1891, 1, 5–6; *Celebration of the Beginning of the Second Century of the American Patent System at Washington City, D.C., April 8, 9, 10, 1891: Proceedings and Addresses* (Washington, DC: Gedney & Roberts, 1892).
2. "Many Men with Brains," *Washington Post,* April 9, 1891, 1.
3. *Celebration,* 24.
4. Ibid., 23.
5. Ibid., 43–127.
6. "Many Men with Brains," 1.
7. Ibid.
8. *Celebration,* 32.
9. Ibid., 424.
10. "New Proofs of the Telephone Fraud," *New York Times,* January 21, 1887, 4.
11. James J. Storrow to John E. Hudson, November 17, 1891, quoted in N. R. Danielian, *A.T.&T.: The Story of Industrial Conquest* (New York: Vanguard Press, 1939), 97.
12. "The Top 100," *Atlantic Monthly* 298 (December 2006): 61–78; "The Greatest American," prod. Jason Raff and Elyse Zaccaro, Discovery Channel and NBC News Productions, June 2005.
13. The source for this original wording is a letter Bell wrote to his father later that day; Alexander Graham Bell to Alexander Melville Bell, March 10, 1876, box 5, Bell Papers.
14. David A. Hounshell, "Elisha Gray and the Telephone: On the Disadvantages of Being an Expert," *Technology and Culture* 16, no. 2 (1975): 133–161; Thomas P. Hughes, *American Genesis: A Century of Invention and Technological Enthusiasm, 1870–1970* (New York: Viking, 1989), 15–16; Michael E. Gorman and W. Bernard Carlson, "Interpreting Invention as a Cognitive Process: The Case of Alexander Graham Bell, Thomas Edison, and the Telephone," *Science, Technology, & Human Values* 15, no. 2 (1990): 131–164; Michael E. Gorman, "Mind in the World: Cognition and Practice in the Invention of the Telephone," *Social Studies of Science* 27, no. 4 (1997): 583–624.
15. See, e.g., Quentin R. Skrabec, *The 100 Most Significant Events in American Business: An Encyclopedia* (Santa Barbara, CA: Greenwood, 2012), 77–80.
16. Daniel J. Boorstin, *The Americans: The Democratic Experience* (New York: Random House, 1973), 58.
17. A. Edward Evenson, *The Telephone Patent Conspiracy of 1876: The Elisha Gray–Alexander Bell Controversy and Its Many Players* (Jefferson, NC: McFarland, 2000); Seth Shulman, *The Telephone Gambit: Chasing Alexander Graham Bell's Secret* (New York: Norton, 2008). Precursors in the genre include Frederick Leland Rhodes,

Beginnings of Telephony (New York: Harper, 1929); William Aitken, *Who Invented the Telephone?* (London: Blackie, 1939); Lewis Coe, *The Telephone and Its Several Inventors* (Jefferson, NC: McFarland, 1995); Burton H. Baker, *The Gray Matter: The Forgotten Story of the Telephone* (St. Joseph, MI: Telepress, 2000).

18. Canonical works include James Willard Hurst, *Law and the Conditions of Freedom in the Nineteenth-Century United States* (Madison: University of Wisconsin Press, 1956); James Willard Hurst, *Law and Economic Growth: The Legal History of the Lumber Industry in Wisconsin, 1836–1915* (Cambridge, MA: Belknap Press of Harvard University Press, 1964); Morton J. Horwitz, *The Transformation of American Law, 1780–1860* (Cambridge, MA: Harvard University Press, 1977).

19. Patent law is absent, e.g., from major studies of judge-made law such as Horwitz, *Transformation of American Law;* P. S. Atiyah, *The Rise and Fall of Freedom of Contract* (Oxford: Clarendon Press, 1979).

20. These include Doron S. Ben-Atar, *Trade Secrets: Intellectual Piracy and the Origins of American Industrial Power* (New Haven, CT: Yale University Press, 2004); Oren Bracha, "Owning Ideas: A History of Intellectual Property in the United States" (SJD diss., Harvard Law School, 2005); B. Zorina Khan, *The Democratization of Invention: Patents and Copyrights in American Economic Development, 1790–1920* (Cambridge: Cambridge University Press, 2005); Steven W. Usselman and Richard R. John, "Patent Politics: Intellectual Property, the Railroad Industry, and the Problem of Monopoly," *Journal of Policy History* 18, no. 1 (2006): 96–125; Catherine L. Fisk, *Working Knowledge: Employee Innovation and the Rise of Corporate Intellectual Property, 1800–1930* (Chapel Hill: University of North Carolina Press, 2009); Kara W. Swanson, "The Emergence of the Professional Patent Practitioner," *Technology and Culture* 50, no. 3 (2009): 519–548; Alain Pottage and Brad Sherman, *Figures of Invention: A History of Modern Patent Law* (Oxford: Oxford University Press, 2010); Adrian Johns, *Piracy: The Intellectual Property Wars from Gutenberg to Gates* (Chicago: University of Chicago Press, 2010); Mario Biagioli, Peter Jaszi, and Martha Woodmansee, eds., *Making and Unmaking Intellectual Property: Creative Production in Legal and Cultural Perspective* (Chicago: University of Chicago Press, 2011); Naomi R. Lamoreaux, Kenneth L. Sokoloff, and Dhanoos Sutthiphisal, "Patent Alchemy: The Market for Technology in U.S. History," *Business History Review* 87, no. 1 (2013): 3–38.

21. Christopher Beauchamp, "Intellectual Property and the Politics of the Telephone Industry in the United States and Britain, 1876–1900" (PhD diss., Cambridge University, 2007), 74.

22. Dan L. Burk and Mark A. Lemley, "Policy Levers in Patent Law," *Virginia Law Review* 89, no. 7 (2003): 1583.

23. "The Bell Telephone Suits," *Scientific American*, February 5, 1887, 80.

24. Abraham Lincoln, "Second Lecture on Discoveries and Inventions," in *The Collected Works of Abraham Lincoln*, ed. Roy P. Basler (New Brunswick, NJ: Rutgers University Press, 1953), 3:363; James Bessen and Michael J. Meurer, *Patent Failure: How Judges, Bureaucrats, and Lawyers Put Innovators at Risk* (Princeton, NJ: Princeton University Press, 2008).

1. Invention in the Lawyers' World

1. Elizabeth S. Kite, *L'Enfant and Washington, 1791–1792: Published and Unpublished Documents Now Brought Together for the First Time* (Baltimore: Johns Hopkins Press, 1929).
2. Kenneth W. Dobyns, *The Patent Office Pony: A History of the Early Patent Office* (Fredericksburg, VA: Sergeant Kirkland's Museum and Historical Society, 1994), chap. 16.
3. John H. Hazelton, *The Declaration of Independence: Its History* (New York: Dodd, Mead, 1906), 290–292; Frederick William True, ed., *A History of the First Half-Century of the National Academy of Sciences, 1863–1913* (Washington, DC: National Academy of Sciences, 1913), 280–282.
4. John Varden, *A Guide for Visitors to the National Gallery, Revised in Accordance with the Instructions of the Commissioner of Patents* (Washington, DC, 1857), http://www.ipmall.info/hosted_resources/patenthistory/poguide.htm.
5. Charles J. Robertson, *Temple of Invention: History of a National Landmark* (Washington, DC: Smithsonian American Art Museum, 2006), 59.
6. Brian Balogh, *A Government Out of Sight: The Mystery of National Authority in Nineteenth-Century America* (Cambridge: Cambridge University Press, 2009).
7. The fee was $30 from 1793 to 1861, and then rose to $35; Patent Act of 1793, Ch. 11, 1 Stat. 318–323 (February 21, 1793); Patent Act of 1861, Ch. 88, 12 Stat. 246 (March 2, 1861).
8. The requirement that patent owners "mark" their patented items originated with the Patent Act of 1842, Ch. 263, 6, 5 Stat. 543–545 (August 29, 1842).
9. B. Zorina Khan, *The Democratization of Invention: Patents and Copyrights in American Economic Development, 1790–1920* (New York: Cambridge University Press, 2005), 182–221.
10. Doron S. Ben-Atar, *Trade Secrets: Intellectual Piracy and the Origins of American Industrial Power* (New Haven, CT: Yale University Press, 2004), 44–77, 159–164.
11. Ibid., 39, 84–86.
12. Christine MacLeod, *Inventing the Industrial Revolution: The English Patent System, 1660–1800* (Cambridge: Cambridge University Press, 1988), 14–19.
13. Oren Bracha, "The Commodification of Patents 1600–1836: How Patents Became Rights and Why We Should Care," *Loyola of Los Angeles Law Review* 38, no. 1 (2004): 202–203; Edward C. Walterscheid, *The Nature of the Intellectual Property Clause: A Study in Historical Perspective* (Buffalo, NY: Hein, 2002), 51–55; H. I. Dutton, *The Patent System and Inventive Activity During the Industrial Revolution, 1750–1852* (Manchester: Manchester University Press, 1984), 75.
14. Turner v. Winter, 99 Eng. Rep. 1274, 1276 (1787).
15. Bruce W. Bugbee, *Genesis of American Patent and Copyright Law* (Washington, DC: Public Affairs Press, 1967), 84–103.
16. James Madison to Thomas Jefferson, October 17, 1788, in *The Republic of Letters: The Correspondence between Thomas Jefferson and James Madison, 1776–1826*, ed. James Morton Smith (New York: Norton, 1995), 1:566.

17. Dotan Oliar, "Making Sense of the Intellectual Property Clause: Promotion of Progress as a Limitation on Congress's Intellectual Property Power," *Georgetown Law Journal* 94 (2006): 1771–1845.
18. U.S. Const., Art. I, Sec. 8.
19. Edward C. Walterscheid, *To Promote the Progress of Useful Arts: American Patent Law and Administration, 1798–1836* (Littleton, CO: Rothman, 1998), 109–110.
20. Patent Act of 1790, Ch. 7, 1 Stat. 109–112 (April 10, 1790).
21. P. J. Federico, "Operation of the Patent Act of 1790," *Journal of the Patent Office Society* 18 (1936): 246; Bracha, "Commodification of Patents," 223.
22. Thomas Jefferson to Hugh Williamson, April 1, 1792, in *The Works of Thomas Jefferson*, ed. Paul Leicester Ford (New York: Putnam, 1904–1905), 6:459.
23. Thomas Jefferson to Thomas Cooper, August 25, 1814, in *The Writings of Thomas Jefferson*, ed. Albert E. Bergh (Washington, DC: Thomas Jefferson Memorial Association, 1907), 14:174.
24. Patent Act of 1793, Ch. 11, 1 Stat. 318–323 (February 21, 1793).
25. Elinor Stearns and David Yerkes, *William Thornton: A Renaissance Man in the Federal City* (Washington, DC: American Institute of Architects Foundation, 1976).
26. Walterscheid, *Promote the Progress*, 253–304.
27. Frank Prager, "The Steamboat Pioneers before the Founding Fathers," *Journal of the Patent Office Society* 37 (1955): 486–522; Frank Prager, "The Steam Boat Interference, 1787–1793," *Journal of the Patent Office Society* 40 (1958): 611–643.
28. Prager, "Steamboat Pioneers," 496–497, 514.
29. Ibid., 496–497.
30. Ibid., 517–518; John Fitch, Petition of May 13, 1789, repr. "Proceedings in Congress during the Years 1789 and 1790, Relating to the First Patent and Copyright Laws," *Journal of the Patent Office Society* 22 (1940): 248.
31. Prager, "Steam Boat Interference," 631–639.
32. Ibid., 640–641.
33. Angela Lakwete, *Inventing the Cotton Gin: Machine and Myth in Antebellum America* (Baltimore: Johns Hopkins University Press, 2003).
34. Denison Olmsted, *Memoir of Eli Whitney, Esq.* (New Haven, CT: Durrie & Peck, 1846), 17; Lakwete, *Inventing the Cotton Gin*.
35. D. A. Tompkins, *The Cotton Gin: The History of Its Invention* (Charlotte, NC: Author, 1901), 14.
36. Olmsted, *Memoir of Eli Whitney*, 27.
37. Patent Act of 1793, Ch. 11, 1 Stat. 318–323 (February 21, 1793), Sec. 5.
38. Olmsted, *Memoir of Eli Whitney*, 26–27; Walterscheid, *Promote the Progress*, 234–235.
39. Patent Act of 1800, Ch. 25, 2 Stat. 37 (April 17, 1800), Sec. 3.
40. Olmsted, *Memoir of Eli Whitney*, 28–46.
41. P. J. Federico, "The Patent Trials of Oliver Evans," pt. 1, *Journal of the Patent Office Society* 27 (1945): 586–589.
42. Ibid., 597.
43. Evans v. Chambers, 8 F. Cas. 837 (C.C.D. Pa. 1807) (No. 4,555).

44. Walterscheid, *Promote the Progress*, 161-163.
45. Act for the Relief of Oliver Evans, 6 Stat. 70 (January 21, 1808); Walterscheid, *Promote the Progress*, 347-349.
46. P. J. Federico, "The Patent Trials of Oliver Evans, " pt. 2, *Journal of the Patent Office Society* 27 (1945): 657-659, 673.
47. Walterscheid, *Promote the Progress*, 350, n. 65.
48. Khan, *Democratization of Invention*, 71.
49. Federico, "Patent Trials of Oliver Evans, " pt. 2; Walterscheid, *Promote the Progress*, 350-354.
50. Federico, "Patent Trials of Oliver Evans," pt. 2, 657.
51. Evans v. Eaton, 20 U.S. (7 Wheat.) 356 (1822).
52. Susan B. Carter, Scott Sigmund Gartner, Michael R. Hanies, Alan M. Olmstead, Richard Sutch, and Gavin Wright, eds., *Historical Statistics of the United States: Millennial Edition* (New York: Cambridge University Press, 2006), table Cg27-37.
53. Kenneth L. Sokoloff, "Inventive Activity in Early Industrial America: Evidence from Patent Records, 1790-1846," *Journal of Economic History* 48, no. 4 (1988): 819-820.
54. Ibid., 828-843.
55. Khan, *Democratization of Invention*, 110-117.
56. David J. Jeremy, *Transatlantic Industrial Revolution: The Diffusion of Textile Technologies between Britain and America, 1790-1830s* (Cambridge, MA: MIT Press, 1981), 184-186.
57. Kenneth L. Sokoloff and B. Zorina Khan, "The Democratization of Invention during Early Industrialization: Evidence from the United States, 1790-1846," *Journal of Economic History* 50, no. 2 (1990): 363-378.
58. William Thornton to Amos Eaton, May 5, 1809, quoted in Walterscheid, *Promote the Progress*, 323.
59. Walterscheid, *Promote the Progress*, 322-324.
60. Thompson v. Haight, 23 F. Cas. 1040, 1041 (C.C.S.D.N.Y. 1826) (No. 13,957).
61. Ibid., 1042.
62. Patent Act of 1836, Ch. 357, 5 Stat. 117 (July 4, 1836); Steven Lubar, "The Transformation of Antebellum Patent Law," *Technology and Culture* 32, no. 4 (1991): 940-942; Walterscheid, *Promote the Progress*, 322-345, 421-432.
63. B. M. Federico, "The Patent Office Fire of 1836," *Journal of the Patent Office Society* 19 (1937): 804-833.
64. Sokoloff, "Inventive Activity," 818-820.
65. Carter et al., *Historical Statistics of the United States*, table Cg27-37.
66. Kara W. Swanson, "The Emergence of the Professional Patent Practitioner," *Technology and Culture* 50, no. 3 (2009): 519-548.
67. Naomi R. Lamoreaux and Kenneth L. Sokoloff, "Intermediaries in the U.S. Market for Technology, 1870-1920," in *Finance, Intermediaries, and Economic Development*, ed. Stanley L. Engerman, Phillip T. Hoffman, Jean-Laurent Rosenthal, and Kenneth L. Sokoloff (Cambridge: Cambridge University Press, 2003), 209-246.
68. Patent Act of 1836, Ch. 357, 5 Stat. 117 (July 4, 1836), Sec. 18.
69. Zebulon Parker, "Sketch of the Invention of Parker's Water Wheel," *Journal of the Franklin Institute* 52, no.1 (1851): 48-50; Edwin T. Layton, "Scientific Technology,

1845–1900: The Hydraulic Turbine and the Origins of American Industrial Research," *Technology and Culture* 20, no. 1 (1979): 68–70.
70. "Controversial—Parker's Water Wheels," *Scientific American*, October 6, 1849, 21.
71. Ibid.; "Woodworth's and Parker's Renewal of Patents," *Scientific American*, April 24, 1852, 251; "Parker and Re-action Water Wheels," *Scientific American*, March 20, 1852, 211.
72. Reporter's note in Parker v. Hatfield, 18 F. Cas. 1127 (C.C.Ohio 1845) (No. 10,736).
73. Reporter's note in Parker v. Stiles, 1 Fish. Pat. Rep. 319 (C.C.Ohio 1849) (No. 10,749).
74. U.S. Circuit Court for the Eastern District of Pennsylvania, Equity Case Files, box 37, NARA Philadelphia.
75. Carolyn C. Cooper, "A Patent Transformation: Woodworking Mechanization in Philadelphia, 1830–1856," in *Early American Technology: Making and Doing Things from the Colonial Era to 1850*, ed. Judith A. McGaw (Chapel Hill: University of North Carolina Press, 1994), 316.
76. "Thomas Blanchard," Report to accompany Bill H.R. 417, April 8, 1834, 23rd Cong., 1st Sess., H.Rep. 397.
77. Carolyn C. Cooper, "Social Construction of Invention through Patent Management: Thomas Blanchard's Woodworking Machinery," *Technology and Culture* 32, no. 4 (1991): 982.
78. Carolyn C. Cooper, *Shaping Invention: Thomas Blanchard's Machinery and Patent Management in Nineteenth-Century America* (New York: Columbia University Press, 1991), 48–54.
79. Brooks v. Fiske, 56 U.S. 212, 224 (1853) (McLean, J., dissenting).
80. Cooper, "Patent Transformation," 293.
81. Report on the Woodworth Patent, July 17, 1852, 32nd Cong., 1st Sess., H.Rep. 156, 3–6.
82. Report on the Administrator of Wm. Woodworth, March 13, 1850, 31st Cong., 1st Sess., H.Rep. 150, 4–5.
83. Richard R. John, *Network Nation: Inventing American Telecommunications* (Cambridge, MA: Belknap Press of Harvard University Press, 2010), 45.
84. Ibid., 35–51.
85. Ibid., 66–68, 74–75.
86. Carl B. Swisher, *The Taney Period, 1836–64* (New York: Macmillan, 1974), 488–504.
87. U.S. Patent RE79, issued to Samuel F. B. Morse for "Improvement in the mode of communicating information by the application of electro-magnetism," January 15, 1846; U.S. Patent RE117, issued to Samuel F. B. Morse for "Improvement in electric telegraphs," June 13, 1848.
88. O'Reilly v. Morse, 56 U.S. (15 How.) 62 (1853).
89. U.S. Patent RE117, claim 8.
90. Cai Guise-Richardson, "Redefining Vulcanization: Charles Goodyear, Patents, and Industrial Control, 1834–1865," *Technology and Culture* 51, no. 2 (2010): 361–370.
91. This measure was a spontaneous initiative of the superintendent of patents in 1813; it received judicial approval from a lower court in 1824 and in the Supreme Court by 1832, and was adopted in the statutes of 1832 and 1836.

92. U.S. Patent 3,633, issued to Charles Goodyear for "Improvement in India-Rubber Fabrics," June 15, 1844; Guise-Richardson, "Redefining Vulcanization," 377-378.
93. Richard Korman, *The Goodyear Story: An Inventor's Obsession and the Struggle for a Rubber Monopoly* (San Francisco: Encounter Books, 2002), 105-106; Guise-Richardson, "Redefining Vulcanization," 360, 375-376.
94. See, e.g., R. C. Grier, *Decision in the Great India Rubber Case of Charles Goodyear vs. Horace H. Day* (New York, 1852).
95. "Decision of the Hon. Joseph Holt, Commissioner of Patents, in the Matter of the Application of Charles Goodyear for the Extension of Letters Patent," quoted in "Charles Goodyear," *North American Review* 101 (1865): 98.
96. David A. Hounshell, *From the American System to Mass Production, 1800-1932: The Development of Manufacturing Technology in the United States* (Baltimore: Johns Hopkins University Press, 1984), 154.
97. Reuben Gold Thwaites, "Cyrus Hall McCormick and the Reaper," *Proceedings of the State Historical Society of Wisconsin* (1908), 247-248.
98. Swisher, *Taney Period*, 505-510; Gordon M. Winder, *The American Reaper: Harvesting Networks and Technology, 1830-1910* (Burlington, VT: Ashgate, 2012), 45.
99. Adam Mossoff, "The Rise and Fall of the First American Patent Thicket: The Sewing Machine War of the 1850s," *Arizona Law Review* 53 (2011): 165-211.
100. Ibid., 193.
101. Parker v. Sears, 18 F. Cas. 1159, 1160 (C.C.E.D. Pa. 1850).
102. Khan, *Democratization of Invention*, 77, 99.
103. See, e.g., Parker v. Haworth, 18 F. Cas. 1135 (C.C.D. Ill. 1848) (on the strict liability of infringers); Parker v. Stiles, 18 F. Cas. 1163 (C.C.D. Ohio 1849) (on infringement by "mechanical equivalents"); Parker v. Hulme, 18 F. Cas. 1138 (C.C.E.D. Pa. 1849) (on protecting the "principle" of an invention).
104. Adam Mossoff, "Who Cares What Thomas Jefferson Thought about Patents? Reevaluating the Patent 'Privilege' in Historical Context," *Cornell Law Review* 92 (2007): 953-1012.
105. Blanchard v. Sprague, 3 F. Cas. 648, 649-650 (C.C.D. Mass., 1839).
106. Ibid., 650.
107. Sloat v. Patton, 1 Fish. Pat. Cas. 154 (E.D. Pa 1852).
108. Lubar, "Transformation of Antebellum Patent Law," 941-942; Mossoff, "Who Cares What Thomas Jefferson Thought," 1000.
109. Brooks v. Fiske, 56 U.S. 212, 214-215 (1853).
110. Winans v. Denmead, 56 U.S. (15 How.) 330 (1853); O'Reilly v. Morse, 56 U.S. 62 (1853).
111. John F. Duffy, "The Festo Decision and the Return of the Supreme Court to the Bar of Patents," *Supreme Court Review* (2002): 289.
112. Andrew Morriss and Craig Nard, "Institutional Choice and Interest Groups in the Development of American Patent Law: 1790-1865," *Supreme Court Economic Review* 19, no. 1 (2011): 143-244.
113. "Colt Patent, &c. &c.," August 3, 1854, 33rd Cong., 1st Sess., H.Rep. 353.
114. Ibid., 4.
115. Ibid., 6-7.

116. "Zebulon and Austin Parker," July 14, 1854, 33rd Cong., 1st Sess., H.Rep. 297.
117. Cooper, "Patent Transformation," 313–315.
118. Patent Act of 1861, Ch. 88, 12 Stat. 246 (March 2, 1861). Extensions of pre-1861 patents continued, however.
119. Robert V. Remini, *Daniel Webster: The Man and His Time* (New York: Norton, 1997), 731–732.
120. Bernard C. Steiner, *Life of Reverdy Johnson* (Baltimore: Norman, Remington, 1914), 36.
121. Robert Henry Parkinson, "The Patent Case That Lifted Lincoln into a Presidential Candidate," *Abraham Lincoln Quarterly* 4 (September 1946): 105–122.
122. Swanson, "Emergence of the Professional Patent Practitioner"; Mason, Fenwick & Lawrence, *Patents for Profit* (Washington, DC, 1907).
123. Robert T. Swaine, *The Cravath Firm and Its Predecessors* (New York: Ad Press, 1946), 1:91–92, 118–120, 152–154; Blatchford, Seward & Griswold Collection, MC 4, MIT Institute Archives and Special Collections, Cambridge, MA.
124. Albert H. Walker, "George Harding," in *Great American Lawyers,* ed. William Draper Lewis (Philadelphia: John C. Winston, 1909), 8:43–87; David McAdam, Henry Bischoff, Jr., Richard H. Clarke, Jackson O. Dykman, Joshua M. Van Cott, and George G. Reynolds, eds., *History of the Bench and Bar of New York* (New York: New York History Company, 1897), 2:213.
125. Carolyn Kinder Carr, *A Brush with History: Paintings from the National Portrait Gallery* (Washington, DC: Smithsonian Institution, 2001), 122.
126. The litigants were Thomas Blanchard, 1788–1864 (inventor of woodworking lathes); Elias Howe, 1819–1867 (sewing machine); William Morton, 1819–1868 (codiscoverer of anesthesia); Samuel Colt, 1814–1862 (firearms); Cyrus McCormick, 1809–1884 (mechanical reaper); Charles Goodyear, 1800–1860 (vulcanized rubber); Samuel Morse, 1791–1872 (electric telegraph); Frederick Sickels, 1819–1895 (cutoff valve crucial to stationary steam engines); Henry Burden, 1791–1871 (horseshoe-making machine and other ironworking); James Bogardus, 1800–1874 (various inventions ranging from engraving machines to cast-iron construction methods); Erastus Bigelow, 1814–1879 (power carpet looms).
127. Carter et al., *Historical Statistics of the United States,* table Cg27–37.
128. Herbert Hovenkamp, *Enterprise and American Law 1836–1937* (Cambridge, MA: Harvard University Press, 1991), 109–110.

2. Acts of Invention

1. James D. Reid, *The Telegraph in America: Its Founders, Promoters and Noted Men* (New York: Derby Brothers, 1879), 596–636; Joel A. Tarr, Thomas Finholt, and David Goodman, "The City and the Telegraph: Urban Telecommunications in the Pre-Telephone Era," *Journal of Urban History* 14, no. 1 (1987): 38–80.
2. Paul Israel, *Edison: A Life of Invention* (New York: Wiley, 1998), 49–65; Paul Israel, *From Machine Shop to Industrial Laboratory: Telegraphy and the Changing Context of American Invention, 1830–1920* (Baltimore: Johns Hopkins University Press, 1992), 124–127, 142–146; Richard R. John, *Network Nation: Inventing American*

Telecommunications (Cambridge, MA: Belknap Press of Harvard University Press, 2010), 149–152.
3. Israel, *Machine Shop to Industrial Laboratory,* 137–138.
4. Item from the *New York Sun,* quoted in "Has Edison Really Accomplished Anything?," *Manufacturer and Builder,* September 1879, 207.
5. John, *Network Nation,* 158–160.
6. David A. Hounshell, "Elisha Gray and the Telephone: On the Disadvantages of Being an Expert," *Technology and Culture* 16, no. 2 (1975): 133–161; Israel, *Edison,* 109.
7. The duplex payment amount appears in President Norvin Green's letterbook, Green to General Wager Swayne, March 19, 1881, Western Union Archives, Smithsonian Institution, Washington, DC. I am grateful to Professor David Hochfelder for supplying me with this reference. On the quadruplex sale, see Israel, *Edison,* 102.
8. Robert V. Bruce, *Bell: Alexander Graham Bell and the Conquest of Solitude* (London: Gollancz, 1973), 92–95.
9. Hounshell, "Elisha Gray and the Telephone"; Michael E. Gorman and W. Bernard Carlson, "Interpreting Invention as a Cognitive Process: The Case of Alexander Graham Bell, Thomas Edison, and the Telephone," *Science, Technology, & Human Values* 15, no. 2 (1990): 131–164.
10. Naomi R. Lamoreaux and Kenneth L. Sokoloff, "Market Trade in Patents and the Rise of a Class of Specialized Inventors in the 19th-Century United States," *American Economic Review* 91, no. 2 (2001): 39–44; Naomi R. Lamoreaux and Kenneth L. Sokoloff, "Inventors, Firms, and the Market for Technology in the Late Nineteenth and Early Twentieth Centuries," in *Learning by Doing in Markets, Firms, and Countries,* ed. Naomi R. Lamoreaux, Daniel M. G. Raff, and Peter Temin (Chicago: University of Chicago Press, 1999), 19–60; Alfred D. Chandler, *The Visible Hand: The Managerial Revolution in American Business* (Cambridge, MA: Belknap Press of Harvard University Press, 1977), 374–375; L. S. Reich, *The Making of American Industrial Research: Science and Business at G.E. and Bell, 1876–1926* (Cambridge: Cambridge University Press, 1985).
11. Bruce Hunt, "'Practice vs. Theory': The British Electrical Debate, 1888–1891," *Isis* 74, no. 3 (1983): 341–355.
12. See generally Israel, *Machine Shop to Industrial Laboratory.*
13. Bruce, *Bell,* 92–94.
14. John, *Network Nation,* 126–128.
15. Rosario J. Tosiello, *The Birth and Early Years of the Bell Telephone System, 1876–1880* (New York: Arno Press, 1979), 8. Thomas Watson later joined the Patent Association, receiving a 10 percent stake.
16. George Crossette, *Founders of the Cosmos Club of Washington, 1878* (Washington, DC: Cosmos Club, 1966), 20; Stanley Harrold, *The Abolitionists and the South, 1831–1861* (Lexington: University Press of Kentucky, 1995), 165.
17. Alexander Graham Bell to Alexander Melville Bell, February 29, 1876, box 5, Bell Papers.
18. A. Edward Evenson, *The Telephone Patent Conspiracy of 1876: The Elisha Gray-Alexander Bell Controversy and Its Many Players* (Jefferson, NC: McFarland, 2000), 43–44.

19. H. Howson to Congressman Thomas A. Jenckes, February 3, 1866, Thomas A. Jenckes Papers (1836–1878), Manuscript Division, Library of Congress, Washington, DC.
20. Evenson, *Telephone Patent Conspiracy*, 43–45.
21. Alexander Graham Bell to Alexander Melville Bell, Eliza Symonds Bell, Carrie Bell, November 23, 1874, box 4, Bell Papers.
22. Gardiner Greene Hubbard to Alexander Graham Bell, August 15, 1874, box 79, Bell Papers.
23. Bruce, *Bell,* 138; Evenson, *Telephone Patent Conspiracy,* 46–47.
24. Tosiello, *Birth and Early Years,* 10; Bruce, *Bell,* 141.
25. Matthew Josephson, *Edison: A Biography* (London: Eyre & Spottiswoode, 1961), 139–144.
26. Alexander Graham Bell to Sarah Fuller, July 1, 1875, box 176, Bell Papers.
27. Gardiner Greene Hubbard to Alexander Graham Bell, January 15, 1876, and Alexander Graham Bell to Gardiner Greene Hubbard, March 14, 1878, box 79, Bell Papers.
28. Gardiner Greene Hubbard to Alexander Graham Bell, January 15, 1876, box 79, Bell Papers.
29. Bruce, *Bell,* 165.
30. The Telephone Cases, 126 U.S. 1 (1888), 567.
31. Hounshell, "Elisha Gray and the Telephone," 148–149, 152–154.
32. The two principal investigative histories are Evenson, *Telephone Patent Conspiracy,* and Seth Shulman, *The Telephone Gambit: Chasing Alexander Graham Bell's Secret* (New York: Norton, 2008).
33. Evenson, *Telephone Patent Conspiracy,* 68–93.
34. Ibid., 80–82.
35. Ibid., 43–47.
36. Alexander Graham Bell, U.S. Patent 174,465, March 7, 1876, for "Improvements in telegraphy."
37. Ibid.
38. Patent Act of 1870, Ch. 230, 16 Stat. 198–217 (July 8, 1870), Sec. 26.
39. Alexander Graham Bell to Mabel Hubbard Bell, January 21, 1879, box 36, Bell Papers.
40. Evenson, *Telephone Patent Conspiracy,* 95–99.
41. Bruce, *Bell,* 188–215.
42. Alexander Graham Bell, U.S. Patent 186,787, January 30, 1877, for "Improvement in electric telegraphy."
43. *American Bell Telephone Company v. Overland Telephone Company of New Jersey et al., Brief for Complainants on Motion for Preliminary Injunction* (Framingham, MA: Clark, 1884), 118–119; argument of J. J. Storrow in the Telephone Cases, 126 U.S. 1 (1888), 317–318.
44. American Bell Tel. Co. v. Brown Tel. & Tel. Co., 58 F. 409 (C.C.N.D. Ill. 1893); American Bell Tel. Co. v. Western Tel. Const. Co., 58 F. 410 (C.C.N.D. Ill. 1893); American Bell Tel. Co. v. McKeesport Tel. Co., 57 F. 661 (C.C.W.D. Pa. 1893).
45. Jeannette Mirsky and Allan Nevins, *The World of Eli Whitney* (New York: Macmillan, 1952), 92–110; Richard Korman, *The Goodyear Story: An Inventor's Obsession and*

the *Struggle for a Rubber Monopoly* (San Francisco: Encounter Books, 2002), 104–105.

46. Christine MacLeod, "The Paradoxes of Patenting: Invention and Its Diffusion in 18th- and 19th-Century Britain, France, and North America," *Technology and Culture* 32, no. 4 (1991): 887; Christine MacLeod, "Strategies for Innovation: The Diffusion of New Technology in Nineteenth-Century British Industry," *Economic History Review* 45, no. 2 (1992): 285-307.
47. Carolyn C. Cooper, *Shaping Invention: Thomas Blanchard's Machinery and Patent Management in Nineteenth-Century America* (New York: Columbia University Press, 1991), 44; MacLeod, "Strategies for Innovation," 295-296.
48. Steven W. Usselman, *Regulating Railroad Innovation: Business, Technology, and Politics in America, 1840–1920* (Cambridge: Cambridge University Press, 2002), 105-107.
49. Harold C. Passer, *The Electrical Manufacturers, 1875–1900: A Study in Competition, Entrepreneurship, Technical Change, and Economic Growth* (Cambridge, MA: Harvard University Press, 1953); Steven W. Usselman, "Patents Purloined: Railroads, Inventors, and the Diffusion of Innovation in 19th-Century America," *Technology and Culture* 32, no. 4 (1991): 1075.
50. John, *Network Nation*, 161-162.
51. Ibid., 163.
52. Bruce, *Bell*, 230-231.
53. George David Smith, *The Anatomy of a Business Strategy: Bell, Western Electric and the Origins of the American Telephone Industry* (Baltimore: Johns Hopkins University Press, 1985), 29-35.
54. Hubbard summarized agency arrangements for the board of the Bell Telephone Company: "Report from Gardiner Greene Hubbard to Stockholders and Directors of the Bell Telephone Company, 1879," Bell Papers.
55. J. J. Storrow to Russell, January 12, 1884, box 1093, AT&T Archives.
56. United States v. United Shoe Machinery of New Jersey, 222 F. 349 (C.C.D. Mass. 1915), 359-360; Blanche Hazard, *The Organization of the Boot and Shoe Industry in Massachusetts before 1875* (Cambridge, MA: Harvard University Press, 1921), 121-122; Ross Thomson, *The Path to Mechanized Shoe Production in the United States* (Chapel Hill: University of North Carolina Press, 1989), 160-163.
57. Smith, *Anatomy of a Business Strategy*, app. C, 163-165.
58. Scott E. Masten and Edward A. Snyder, "United States versus United Shoe Machinery Corporation: On the Merits," *Journal of Law and Economics* 36, no. 1 (1993): 35-46.
59. Frank Thomas, "The Politics of Growth: The German Telephone System," in *The Development of Large Technical Systems*, ed. Renate Mayntz and Thomas P. Hughes (Boulder, CO: Westview Press, 1988), 183-187; Wilfried Feldenkirchen, *Werner von Siemens: Inventor and International Entrepreneur* (Columbus: Ohio State University Press, 1994), 110.
60. Israel, *Machine Shop to Industrial Laboratory*, 136-138; Israel, *Edison*, 62-65.
61. Tosiello, *Birth and Early Years*, 225-228.

62. Alexander Graham Bell to Gardiner Greene Hubbard, July 28, 1880, box 79, Bell Papers; Tosiello, *Birth and Early Years*, 158–159, 185.
63. Robert W. Garnet, *The Telephone Enterprise: The Evolution of the Bell System's Horizontal Structure, 1876–1909* (Baltimore: Johns Hopkins University Press, 1985), 18–20, 26–27, 38–43; Smith, *Anatomy of a Business Strategy*, 44–49; Robert MacDougall, *The People's Network: The Political Economy of the Telephone in the Gilded Age* (Philadelphia: University of Pennsylvania Press, 2013), 71–73.
64. Tosiello, *Birth and Early Years*, 457.
65. Ibid., 409; Garnet, *Telephone Enterprise*, 47, 52.
66. United States v. American Bell Telephone Company, 167 U.S. 224, 243 (1897); Charles Cheever to Gardiner Greene Hubbard, February 28, 1878, box 1205, AT&T Archives; Charles H. Aldrich, *The American Bell Telephone Monopoly and the Pending Legislation in Its Interest: A Memorial to the Fifty-Third Congress* (Chicago: privately printed, 1894), 7.
67. James J. Storrow and Chauncey Smith, "Briefs for Alexander Graham Bell and Francis Blake, Jr." (1881), TI4, Edison Papers.
68. Tosiello, *Birth and Early Years*, 349–353, 452–468; Garnet, *Telephone Enterprise*, 46–49.
69. John, *Network Nation*, 164–170.
70. David Hochfelder, "Constructing an Industrial Divide: Western Union, AT&T, and the Federal Government, 1876–1971," *Business History Review* 76, no. 4 (2002): 713–715; Kenneth Lipartito, *The Bell System and Regional Business: The Telephone in the South, 1877–1920* (Baltimore: Johns Hopkins University Press, 1989), 49–50.
71. Lipartito, *Bell System and Regional Business*, 48–51.
72. "A List of Western Union Telephone, Etc. Patents for All Time," box 1005, AT&T Archives.
73. Tosiello, *Birth and Early Years*, 486.
74. Garnet, *Telephone Enterprise*, 58–61; Smith, *Anatomy of a Business Strategy*, 104–107.

3. The Telephone Cases

1. This collective term was (and still is) used for the Supreme Court's consolidated decision on all the main challenges to Bell's patent. See the Telephone Cases, 126 U.S. 1 (1888).
2. "The Bell Telephone Suits," *Scientific American*, February 5, 1887, 80.
3. *Proceedings of the Bench and Bar of the Supreme Court of the United States in Memoriam Morrison R. Waite* (Washington, DC: Government Printing Office, 1888), 37.
4. The Civil Rights Cases, 109 U.S. 3 (1883); Munn v. Illinois, 94 U.S. 113 (1877).
5. Paul David, "Heroes, Herds and Hysteresis in Technological History: Thomas Edison and 'The Battle of the Systems' Reconsidered," *Industrial and Corporate Change* 1, no. 1 (1992): 129–180.
6. Joel Mokyr, *The Lever of Riches: Technological Creativity and Economic Progress* (Oxford: Oxford University Press, 1990), 13–14.

7. Ibid., 292.
8. Patent Act of 1836, Ch. 357, 5 Stat. 117 (July 4, 1836), Sec. 6.
9. Patent Act of 1870, Ch. 230, 16 Stat. 198–217 (July 8, 1870), Sec. 26.
10. These approaches are typically referred to respectively as "central claiming" and "peripheral claiming." Jeanne Fromer, "Claiming Intellectual Property," *University of Chicago Law Review* 76 (2009): 731–742.
11. Alain Pottage and Brad Sherman, *Figures of Invention: A History of Modern Patent Law* (Oxford: Oxford University Press, 2010), 127–141.
12. Warren T. Jessup, "The Doctrine of Equivalents," *Journal of the Patent Office Society* 54 (1972): 248; Suzanne Scotchmer, "Standing on the Shoulders of Giants: Cumulative Research and the Patent Law," *Journal of Economic Perspectives* 5, no.1 (1991): 30, n. 2.
13. William K. Townsend, "Patents," in *Two Centuries' Growth of American Law, 1701–1901,* ed. Members of the Faculty of the Yale Law School (New York: Charles Scribner's Sons, 1901), 406; Albert H. Walker, *Text-Book of the Patent Laws of the United States of America* (New York: Strouse, 1883), 262–265; Simon G. Croswell, "Infringement Cases in Patent Law," *Harvard Law Review* 3, no. 5 (1889): 206–212.
14. Chicago & N.W. Railway Co. v. Sayles, 97 U.S. 554, 556–557 (1878).
15. Figures calculated from the U.S. Patent Office, *Annual Report of the Commissioner of Patents for the Year 1892* (Washington, DC: Government Printing Office, 1893), xii.
16. Oren Bracha, "Geniuses and Owners: The Construction of Inventors and the Emergence of American Intellectual Property," in *Transformations in American Legal History: Essays in Honor of Professor Morton J. Horwitz,* ed. Daniel W. Hamilton and Alfred L. Brophy (Cambridge, MA: Harvard University Press, 2009), 369–390.
17. Martha Woodmansee, "The Genius and the Copyright: Economic and Legal Conditions of the Emergence of the 'Author,'" *Eighteenth-Century Studies* 17, no. 4 (1984): 425–448; Mark Rose, *Authors and Owners: The Invention of Copyright* (Cambridge, MA: Harvard University Press, 1993); Oren Bracha, "The Ideology of Authorship Revisited: Authors, Markets, and Liberal Values in Early American Copyright," *Yale Law Journal* 118 (2008): 186–271.
18. Brooke Hindle, *Emulation and Invention* (New York: Norton, 1983), 128.
19. Thomas P. Hughes, *American Genesis: A Century of Invention and Technological Enthusiasm, 1870–1970* (New York: Viking, 1989), 15–16; Merritt Roe Smith, "Technological Determinism in American Culture," in *Does Technology Drive History? The Dilemma of Technological Determinism,* ed. Merritt Roe Smith and Leo Marx (Cambridge, MA: MIT Press, 1994), 5–8.
20. Catherine Fisk, *Working Knowledge: Employee Innovation and the Rise of Corporate Intellectual Property, 1800–1930* (Chapel Hill: University of North Carolina Press, 2009), 109–126.
21. Chicago & N.W. Railway Co. v. Sayles, 556.
22. William W. Fisher, Morton J. Horwitz, and Thomas A. Reed, eds., *American Legal Realism* (New York: Oxford University Press, 1993).
23. Alexander Graham Bell to Mabel Hubbard Bell, January 26, 1879, box 36, Bell Papers.

24. "Smith, Chauncey," *Dictionary of American Biography* (New York: Scribner's, 1928), 17:253–254; Robert V. Bruce, *Bell: Alexander Graham Bell and the Conquest of Solitude* (London: Gollancz, 1973), 268.
25. "Storrow, James Jackson," *Dictionary of American Biography,* 18:99–100; Herbert Casson, *The History of the Telephone* (Chicago: McClurg, 1910), 101–104; Bruce, *Bell,* 267–268.
26. "Colt Patent, &c. &c.," August 3, 1854, 33rd Cong., 1st Sess., H.Rep. 353, 3–5; "Death of E. N. Dickerson; Close of the Life of a Noted Lawyer," *New York Times,* December 13, 1889.
27. Alexander Graham Bell to Mabel Hubbard Bell, January 22, 1879, box 36, Bell Papers. Emphasis in original.
28. Alexander Graham Bell, U.S. Patent 174,465, March 7, 1876, for "Improvements in telegraphy."
29. Alexander Graham Bell to Mabel Hubbard Bell, January 26, 1879, box 36, Bell Papers.
30. Ibid.
31. *In the U.S. Circuit Court for the District of Massachusetts, Bell Telephone Company et al. v. Peter A. Dowd. Pleadings, Evidence and Exhibits* (Boston, 1880).
32. On Gray, see David A. Hounshell, "Elisha Gray and the Telephone: On the Disadvantages of Being an Expert," *Technology and Culture* 16, no. 2 (1975): 133–161.
33. Rosario J. Tosiello, *The Birth and Early Years of the Bell Telephone System, 1876–1880* (New York: Arno Press, 1979), 225–228.
34. David Hochfelder, "Constructing an Industrial Divide: Western Union, AT&T, and the Federal Government, 1876–1971," *Business History Review* 76, no. 4 (2002): 713–715; Kenneth Lipartito, *The Bell System and Regional Business: The Telephone in the South, 1877–1920* (Baltimore: Johns Hopkins University Press, 1989), 49–50.
35. American Bell Telephone Company v. Spencer, 8 F. 509 (C.C.D. Mass. 1881); Frederick Leland Rhodes, *Beginnings of Telephony* (New York: Harper, 1929), 208–209.
36. American Bell Telephone Company v. Spencer, 509–510.
37. Ibid., 511.
38. Ibid., 512.
39. "Important Telephone Decision," *Scientific American,* July 16, 1881, 32. Emphasis in original.
40. American Bell Telephone Company v. Dolbear, 15 F. 448 (C.C.D. Mass. 1883).
41. Ibid., 453.
42. George Carroll, "Process Patents Involving Principles of Nature," *Yale Law Journal* 19, no. 3 (1910): 162–179.
43. Oren Bracha, "Owning Ideas: A History of Intellectual Property in the United States" (SJD diss., Harvard Law School, 2005), 448–488.
44. O'Reilly v. Morse, 56 U.S. (15 How.) 62 (1854).
45. American Bell Telephone Company v. Dolbear, 15 F. 448, 449 (C.C.D. Mass. 1883); Bracha, "Owning Ideas," 466.
46. Cochrane v. Deener, 94 U.S. 780, 787 (1876); Tilghman v. Proctor, 102 U.S. 707 (1880).
47. Tilghman v. Proctor, 708.

48. Ibid., 723.
49. Ibid., 726.
50. American Bell Telephone Company v. Spencer, 511–512; American Bell Telephone Company v. Dolbear, 449. Justice Gray had not been a member of the Supreme Court at the time of the *Tilghman* decision.
51. "The Reis Telephones," *Scientific American,* August 22, 1885, 113; "Bell Telephone Suits," 80.
52. Lawrence M. Friedman, *A History of American Law* (New York: Simon and Schuster, 1973), 380; Steven W. Usselman, *Regulating Railroad Innovation: Business, Technology, and Politics in America, 1840–1920* (Cambridge: Cambridge University Press, 2002), 170–171.
53. "Bell Telephone Suits," 80. E.g., Western Union had secured reissue of the Stearns duplex twice: first just one month after purchasing it in 1873, then again in 1882. "Fighting for the Stearns Duplex," *Electrical World,* September 19, 1885, 118–119. See also Kendall J. Dood, "Pursuing the Essence of Inventions: Reissuing Patents in the 19th Century," *Technology and Culture* 32, no. 4 (1991): 1004–1008.
54. The decisive case was Miller v. Bridgeport Brass Co., 104 U.S. 350 (1881). See also Dood, "Pursuing the Essence of Inventions," 1015–1016.
55. Townsend, "Patents," 396.
56. Hotchkiss v. Greenwood, 52 U.S. 248, 267 (1850); B. Zorina Khan, "Property Rights and Patent Litigation in Early Nineteenth-Century America," *Journal of Economic History* 55, no. 1 (1995): 72, n. 25.
57. Atlantic Works v. Brady, 107 U.S. 192, 200 (1883).
58. James W. Ely, *The Guardian of Every Other Right: A Constitutional History of Property Rights,* 2nd ed. (Oxford: Oxford University Press, 1998), 87–93.
59. Arthur P. Greeley, *Foreign Patent and Trademark Laws: A Comparative Study* (Washington, DC: John Byrne, 1899).
60. American Bell Telephone Company v. People's Telephone Company, 22 F. 309, 326 (C.C.S.D.N.Y. 1884).
61. American Bell Telephone Company v. American Cushman Telephone Company, 35 F. 734 (C.C.N.D. Ill. 1888).
62. American Bell Telephone Company v. Globe Telephone Company, 31 F. 729, 733–734 (C.C.S.D.N.Y. 1887). See also Basilio Catania, "Antonio Meucci: Telephone Pioneer," *Bulletin of Science, Technology and Society* 21, no. 1 (2001): 55–76; Basilio Catania, "Antonio Meucci, Inventor of the Telephone: Unearthing the Legal and Scientific Proofs," *Bulletin of Science, Technology and Society* 24, no. 2 (2004): 115–137.
63. Basilio Catania, "The U.S. Government Versus Alexander Graham Bell: An Important Acknowledgment for Antonio Meucci," *Bulletin of Science, Technology and Society* 22, no. 6 (2002): 426–442, n. 4.
64. Silvanus Thompson, *Philipp Reis: Inventor of the Telephone* (London: E. & F. N. Spon, 1883); see also the collection of articles by pro-Reis American scientists reprinted in the *Telegraphic Journal and Electrical Review,* January 22 and 29, 1886.
65. Thompson, *Philipp Reis,* 47–48; American Bell Telephone Company v. Molecular Telephone Company, 32 F. 214, 217–218 (C.C.S.D.N.Y. 1885); A. Edward Evenson,

The Telephone Patent Conspiracy of 1876: The Elisha Gray–Alexander Bell Controversy and Its Many Players (Jefferson, NC: McFarland, 2000), 158.

66. See, e.g., "The Antecedents of the Bell Telephone," *Scientific American*, August 6, 1881, 83–84; *In the U.S. Circuit Court for the District of New Jersey, American Bell v. Ghegan. Affidavits of Defendant*, 30; Warren J. Harder, *Daniel Drawbaugh: The Edison of the Cumberland Valley* (Philadelphia: University of Pennsylvania Press, 1960), 70.

67. "The Problem of the Telephone," *Scientific American*, February 17, 1883, 96.

68. "The Course of Telephone Litigation," *Electrical World*, February 2, 1884, 36; "The Directorate of the Globe Company," *Electrical World*, April 5, 1884, 115; "The Telephone Cases in America," *Telegraphic Journal and Electrical Review*, January 24, 1885, 82; "The Telephone," *Western Electrician*, January 21, 1888, 35; Catania, "U.S. Government versus Alexander Graham Bell," 428.

69. *In the U.S. Circuit Court for the Southern District of New York, American Bell Telephone Company v. The People's Telephone Company et al. Evidence for Complainants* (Boston, 1882), 1:85–104; *In the U.S. Circuit Court for the Southern District of New York, American Bell Telephone Company v. The People's Telephone Company et al., Appendix to Complainants' Briefs* (Boston, 1882), 162–165; The Telephone Cases, 126 U.S. 1, 547–552, 561 (1888).

70. American Bell Telephone Company v. Molecular Telephone Company, 32 F. 214 (C.C.S.D.N.Y 1885); Thomas Lockwood, "Memorandum Relating to the Litigation of the Bell Patents," undated typescript, 37, 40, box 1056, AT&T Archives; Rhodes, *Beginnings of Telephony*, 211–213.

71. U.S. Congress, House, Select Committee on Pan-Electric Telephone Stock, *Report of the Minority* (1886), 49th Cong., 1st Sess., H. Rpt. no. 3142, 73, 84; U.S. Congress, House, Committee on Pan-Electric Telephone Company, *Testimony Taken by Committee on Pan Electric Telephone Company*, 49th Cong., 1st Sess., 1886, 45–51; Rhodes, *Beginnings of Telephony*, 217–218.

72. "The Great Telephone Suit," *Scientific American*, October 4, 1884, 208–209.

73. *In the U.S. Circuit Court for the District of New Jersey, American Bell Telephone Company v. John J. Ghegan. Complainants' Moving Papers on Motion for Preliminary Injunction* (Boston, 1882), 9. On the careers of Harding and Edmunds, see Albert H. Walker, "George Harding," in *Great American Lawyers*, ed. William Draper Lewis (Philadelphia: John C. Winston, 1909), 8:43–87; Samuel B. Hand, "Edmunds, George Franklin (1828–1919)," in *American National Biography*, ed. John A. Garraty and Mark C. Carnes (New York: Oxford University Press, 1999), 7:320–321.

74. Robert Bolt, "Donald McDonald Dickinson," in Garraty and Carnes, *American National Biography*, 6:561–562.

75. Donald Grier Stephenson, *The Waite Court: Justices, Rulings, and Legacy* (Santa Barbara, CA: ABC-CLIO, 2003), 32.

76. These details are from the following *New York Times* reports on the trial: "Talking About the Telephone," September 23, 1884, 8; "Fighting for the Telephone," September 24, 1884, 8; "The Telephone Claimants," September 25, 1884, 8; "Pleading for Mr. Drawbaugh," September 26, 1884, 8; "Drawbaugh's Claim," September 27, 1884, 8;

"The Telephone Suits," October 1, 1884, 8; "The Telephone Arguments Closed," October 3, 1884, 2.
77. American Bell Telephone Company v. People's Telephone Company, 327.
78. American Bell Telephone Company v. Molecular Telephone Company; American Bell Telephone Company v. National Improved Telephone Company, 27 F. 663 (C.C.E.D. La. 1886); Rhodes, *Beginnings of Telephony*, 212–219.
79. "The Legal Effect of the Wallace Decision," *Telegraphic Journal and Electrical Review*, January 10, 1885, 34–35; Affidavit of W. Van Benthuysen, November 13, 1885, in U.S. Department of the Interior, *The Telephone Case: Record* (Washington, DC: Department of the Interior, 1885) (hereafter Interior Department, *Telephone Case*), 178; Lockwood, "Memorandum Relating to the Litigation of the Bell Patents," 44–45.
80. Goodyear Dental Vulcanite Company v. Willis, 10 F. Cas. 754, 756 (C.C. Mich. 1874); Rhodes, *Beginnings of Telephony*, 212–219.
81. U.S. Senate, "Arguments before the Committee on Patents of the Senate and House of Representatives in Support of and Suggesting Amendments to the Bills (S. no. 300 and H.R. 1620) to Amend the Statutes in Relation to Patents," 45th Cong., 2d Sess., misc. doc. no. 50, p. 140; Steven Lubar, "The Transformation of Antebellum Patent Law," *Technology and Culture* 32, no. 4 (1991): 954–958.
82. Khan, "Property Rights and Patent Litigation"; Naomi R. Lamoreaux and Kenneth L. Sokoloff, "Inventors, Firms, and the Market for Technology in the Late Nineteenth and Early Twentieth Centuries," in *Learning by Doing in Markets, Firms, and Countries*, ed. Naomi R. Lamoreaux, Daniel M. G. Raff, and Peter Temin (Chicago: University of Chicago Press, 1999), 19–60.
83. Edward A. Purcell, *Litigation and Inequality: Federal Diversity Jurisdiction in Industrial America, 1870–1958* (New York: Oxford University Press, 1992).
84. American Bell Telephone Company v. Molecular Telephone Company, 216; "Another Telephone Decision," *Scientific American*, July 18, 1885, 32; American Bell Telephone Company v. National Improved Telephone Company, 664–665.
85. R. L. Mahon, "The Telephone in Chicago, Brief Based on Materials Collected by Historical Committee," ms. c. 1951, box 447 04 02 / 01, AT&T Archives; Resolution of New York Board of Trade and Transportation, October 14, 1885, in Interior Department, *Telephone Case*, 24c–24d; "The Telephone in New England," *Electrical World*, May 16, 1885, 198–199.
86. *Electrical World*, vols. 5 and 6, weekly listings of stock prices. Year-beginning and year-ending prices taken January 3 and December 26, 1885.
87. W. H. Forbes to J. J. Storrow, September 13, 1884, quoted in Arthur S. Pier, *Forbes: Telephone Pioneer* (New York: Dodd, Mead, 1953), 149.
88. See generally the journal *Western Electrician*, vols. 2 and 3, especially "The Telephone," *Western Electrician*, January 21, 1888, 35; R. L. Mahon, "The Telephone in Chicago."
89. "A Million on Paper for $100 Cash," *Electrical World*, August 8, 1885, 59.
90. Thomas Lockwood, "Memorandum Relating to the Litigation of the Bell Patents," 21, 40.
91. "Bell Telephone Suits," 80; Thomas Lockwood, "Memorandum relating to the Litigation of the Bell Patents," 62.

92. Editorial, *Electrical World*, March 14, 1885, 101.
93. "Bell and the Bench," *New York Herald*, December 2, 1886, 3; "Remarkable Revelations," *New York Herald*, December 2, 1886, 6; "Hello! Hello!," *New York Herald*, December 3, 1886, 3; "The Bar and the Bell," *New York Herald*, December 3, 1886, 3.
94. Richard White, "Information, Markets, and Corruption: Transcontinental Railroads in the Gilded Age," *Journal of American History* 90, no. 1 (2003): 19–43.
95. Perkins's allegiances are far from clear. In one of his letters, he appeared to seek a quid pro quo from the attorney general in a dispute between himself and the federal court in Boston. John M. Perkins to Solicitor General Jenks, January 3, 1888, RG60.3.2, year file 1885 / 6921, box 137, folder 4, NARA. Perkins's earlier letters to Garland, dated July 1, 3, 6, and 8, 1886, are at RG60.3.2, year file 1885 / 6921, box 136, folder 2, NARA; "Hello! Hello!," 3.
96. The proprietors and editors of the main newspapers involved gave testimony to the congressional committee investigating the Pan-Electric affair: see Committee on Pan-Electric Telephone Company, *Testimony Taken by Committee*, testimony of Joseph Pulitzer, Stilson Hutchins, Charles Dana, George Jones, and S. H. Clark.
97. Ted Curtis Smythe, *The Gilded Age Press, 1865–1900* (Westport, CT: Praeger, 2003).
98. "An Odious Monopoly," *New York Times*, November 20, 1886, 4; John, *Network Nation*, 207.
99. Committee on Pan-Electric Telephone Company, *Testimony Taken by Committee*, testimony of George Jones, 886.
100. The Telephone Cases, 232–242, 246–248.
101. Ibid., 251–257.
102. Letters of W. C. Barney reporting on oral arguments, Letters Received, General Records of the Department of Justice, 1849–1989, RG 60.3.2, year file 1885 / 6921, box 138, folder 1, NARA; "Bell Telephone Suits," 80.
103. The Telephone Cases, 570.
104. The Telephone Cases, 163–212.
105. Ibid., 329–389.
106. Ibid., 394.
107. *Electrical World*, February 19, 1887, 87; *Boston Evening Transcript*, quoted in "Drawbaugh Stock," *Electrical World*, March 5, 1887, 125.
108. The Telephone Cases, 573 (Bradley, J., dissenting).
109. Ibid., 576.
110. See, e.g., Edward A. Purcell, "The Particularly Dubious Case of Hans v. Louisiana: An Essay on Law, Race, History, and 'Federal Courts,'" *North Carolina Law Review* 81 (2003): 1927–2059.
111. Christopher Beauchamp, "Intellectual Property and the Politics of the Telephone Industry in the United States and Britain, 1876–1900" (PhD diss., Cambridge University, 2007), 78. Overall, Field dissented 233 times during his tenure on the Court, Harlan 380; Friedman, *History of American Law*, 331.
112. Joseph Bradley, "Bell Telephone Patent Case in Supreme Court," 1887, box 13, folder 6, Joseph Bradley Papers.

113. Steven W. Usselman and Richard R. John, "Patent Politics: Intellectual Property, the Railroad Industry, and the Problem of Monopoly," *Journal of Policy History* 18, no. 1 (2006): 117.
114. Swain Turbine & Mfg. Co. v. Ladd, 102 U.S. 408, 411 (1880).
115. Tilghman v. Proctor; see also Atlantic Works v. Brady, 107 U.S. 192 (1883).
116. Joseph Bradley, "Bell Telephone Patent Case in Supreme Court."
117. Webster Loom Co. v. Higgins, 105 U.S. 580, 595 (1881); Atlantic Works v. Brady, 203; Reiter v. Jones & Laughlin Ltd., 35 F. 421, 423 (C.C.W.D. Pa. 1888).
118. Townsend, "Patents," 413; William C. Robinson, *The Law of Patents for Useful Inventions*, 3 vols. (Boston: Little, Brown, 1890), 1:v.

4. THE UNITED STATES VERSUS BELL

1. Arthur P. Greeley, *Foreign Patent and Trademark Laws: A Comparative Study* (Washington, DC: John Byrne, 1899), 83–88.
2. Ibid., 81; B. Zorina Khan, "Property Rights and Patent Litigation in Early Nineteenth-Century America," *Journal of Economic History* 55, no. 1 (1995): 73–74.
3. A. Edward Evenson, *The Telephone Patent Conspiracy of 1876: The Elisha Gray-Alexander Bell Controversy and Its Many Players* (Jefferson, NC: McFarland, 2000); Seth Shulman, *The Telephone Gambit: Chasing Alexander Graham Bell's Secret* (New York: Norton, 2008).
4. Watson van Benthuysen to Augustus Garland, July 12, 1885, repr. in U.S. Department of the Interior, *The Telephone Case: Record* (Washington, DC: Department of the Interior, 1885) (hereafter Interior Department, *Telephone Case*), 1–2; Memorial of McCorry, van Benthuysen, Huntington, Beckwith, and Gantt, August 26, 1885, in Interior Department, *Telephone Case*, 7–15.
5. Wilber's affidavit is reproduced at Evenson, *Telephone Patent Conspiracy*, 167–171.
6. The Telephone Cases, 126 U.S. 1, 232–243 (1888).
7. U.S. House of Representatives, Select Committee on Pan-Electric Telephone Stock, *Report of the Minority* (1886), 49th Cong., 1st Sess., H.Rep. 3142, 90.
8. Ibid., 73–80; Richard John, *Network Nation: Inventing American Telecommunications* (Cambridge, MA: Harvard University Press, 2010), 205–207.
9. John Brooks, *Telephone: The First Hundred Years* (New York: Harper & Row, 1976), 88–89; Kenneth Lipartito, *The Bell System and Regional Business: The Telephone in the South, 1877–1920* (Baltimore: Johns Hopkins University Press, 1989), 82–84.
10. Mowry v. Whitney, 81 U.S. 434 (1871).
11. United States v. American Bell Telephone Company, 128 U.S. 315, 337–339 (1888).
12. Mowry v. Whitney, 441.
13. Attorney General v. Rumford Chemical Works, 32 F. 608 (C.C.D.R.I. 1876); United States v. Frazer, 22 F. 106 (N.D. Ill. 1884); Opinion of Judge Baxter in United States v. Curry (1879), not reported but discussed in a letter from Richards, U.S. attorney for the Southern District of Ohio, to Attorney General Charles Devens, July 25, 1879, RG60.3.2, year file 1885 / 6921, box 135, NARA.
14. United States v. Gunning, 18 F. 511 (C.C.S.D.N.Y 1883).
15. United States v. Colgate, 32 F. 624, 624 (C.C.S.D.N.Y 1884).

16. United States v. Frazer, 107.
17. Report of the Commissioner of Patents on proceedings relating to *Hecker v. Rumford Chemical Works*, April 1875, and briefs in the case of Attorney General v. Rumford Chemical Works, RG60.3.2, year file 1885 / 6921, box 136, NARA; Attorney General v. Rumford Chemical Works.
18. The circumstances of the cases and of the lumbermen's response are described in "The Woodbury Patent," *Scientific American,* January 9, 1875, 16; "The Woodbury Patent," *Scientific American,* September 18, 1875, 176. Details of the government's involvement appear in the letter of J. Drew (counsel for L. Gould) to Attorney General Pierrepont, May 29, 1875, RG60.3.2, year file 1885 / 6921, box 136, NARA. The Supreme Court decided against Woodbury in Woodbury Patent Planing-Machine Company v. Keith, 101 U.S. 479 (1879).
19. John J. McLaurin, *Sketches in Crude Oil: Some Accidents and Incidents of the Petroleum Development in All Parts of the Globe* (Harrisburg, PA: Author, 1896), 333–336.
20. "Statement of General Duncan Walker in reference to the 'Roberts Torpedo' and 'Barbed Wire' cases," in Interior Department, *Telephone Case,* 406–420.
21. Earl W. Hayter, "An Iowa Farmers' Protective Association: A Barbed Wire Patent Protest Movement," *Iowa Journal of History and Politics* 37, no. 4 (1939): 352; J. M. McFadden, "Monopoly in Barbed Wire: The Formation of the American Steel and Wire Company," *Business History Review* 52, no. 4 (1978): 466–470.
22. Hans L. Trefousse, "Benjamin Franklin Butler," in *American National Biography,* ed. John A. Garraty and Mark C. Carnes (New York: Oxford University Press, 1999), 4:91–93; Hayter, "Iowa Farmers' Protective Association," 346.
23. "Statement of General Duncan Walker in reference to the 'Roberts Torpedo' and 'Barbed Wire' cases," in Interior Department, *Telephone Case,* 418–420; Mr. Humphreys to Solicitor General Jenks, January 14, 1887, RG60.3.2, year file 1885 / 6921, box 137, NARA.
24. Earl W. Hayter, "The Patent System and Agrarian Discontent, 1875–1888," *Mississippi Valley Historical Review* 34, no. 1 (1947): 59–82; Steven W. Usselman, *Regulating Railroad Innovation: Business, Technology, and Politics in America, 1840–1920* (Cambridge: Cambridge University Press, 2002), 147–148.
25. See U.S. Senate, Committee on Patents, Patent Infringements and Practice in Patent Suits (unpublished hearing, 1884), 48th Cong., 1st Sess., SPat-T.1; Usselman, *Regulating Railroad Innovation,* 148–149.
26. Usselman, *Regulating Railroad Innovation,* 146–153.
27. Chauncey Smith, "A Century of Patent Law," *Quarterly Journal of Economics* 5, no. 1 (1890): 58–59.
28. C. A. Brown, "Revision of the Patent Law," *Western Electrician,* January 21, 1888, 31; "Convention of the National Electric Light Association at Pittsburgh," *Western Electrician,* February 25, 1888, 89–92; "Reform of the Patent System," *Electrical World,* April 14, 1888, 186.
29. The three patent-cancellation bills were H.R. 6456 reported in 1880, H.R. 6512 reported 1882, and H.R. 3036 reported 1884: all copied in U.S. House of Representatives, Committee on Pan-Electric Telephone Company, *Testimony Taken by Commit-*

tee on Pan Electric Telephone Company (published hearing, 1886), 49th Cong., 1st Sess., H. Misc. Doc. v. 19, no. 355, 694–697.
30. Richard Korman, *The Goodyear Story: An Inventor's Obsession and the Struggle for a Rubber Monopoly* (San Francisco: Encounter Books, 2002), 122.
31. Resolution of New York Board of Trade and Transportation, October 14, 1885, Interior Department, *Telephone Case*, 24c–24d.
32. Secretary of the Interior L. Q. C. Lamar to Solicitor General John Goode, January 14, 1886, RG60.3.2, year file 1885 / 6921, box 136, NARA.
33. Committee on Pan-Electric Telephone Company, *Testimony Taken by Committee*, 697–699.
34. "The Bell Telephone before the Supreme Court," *Scientific American,* February 19, 1887, 113.
35. Grosvenor P. Lowrey to Allen G. Thurman, April 9, 1886, RG60.3.2, year file 1885 / 6921, box 135, NARA.
36. "Affidavit Wilbur Dies in Denver," *Electrical World,* September 7, 1889, 175.
37. Eppa Hunton and Jeff. Chandler to Solicitor General John Goode, undated 1886, RG60.3.2, year file 1885 / 6921, box 136, NARA.
38. Grosvenor P. Lowrey to Solicitor General John Goode, June 23, 1886, RG60.3.2, year file 1885 / 6921, box 135, NARA.
39. Grosvenor P. Lowrey to Solicitor General John Goode, February 9, 1886, RG60.3.2, year file 1885 / 6921, box 136, NARA.
40. H. P. McIntosh to Solicitor General John Goode, February 11, 1886, RG60.3.2, year file 1885 / 6921, box 135, NARA.
41. United States Attorney Richards to Attorney General Charles Devens, July 25, 1879, RG60.3.2, year file 1885 / 6921, box 135, NARA; Committee on Pan-Electric Telephone Company, *Testimony Taken by Committee,* 458–459.
42. United States v. American Bell Telephone Company, 29 F. 17 (C.C.S.D. Ohio 1886).
43. United States v. American Bell Telephone Company, 32 F. 591 (C.C.D. Mass. 1887).
44. United States v. American Bell Telephone Company, 128 U.S. 315, 335 (1888).
45. Ibid., 367.
46. See also *Mahn v. Harwood,* 112 U.S. 354, 364–365 (1884) (Miller, J., dissenting).
47. Grosvenor P. Lowrey to Attorney General William Miller, January 26, 1890, RG60.3.2, year file 1885 / 6921, box 138, NARA.
48. Homer Cummings and Carl McFarland, *Federal Justice: Chapters in the History of Justice and the Federal Executive* (New York: Macmillan, 1937), 304–305.
49. Parker C. Chandler to unidentified government attorney, January 20, 1890, RG60.3.2, year file 1885 / 6921, box 138, NARA.
50. Memorandum from Charles S. Whitman to Attorney General William Miller, September 26, 1893; Printed motion for extension to collect evidence, filed December 11, 1895: both in RG60.3.2, year file 1885 / 6921, box 139, NARA.
51. Timothy R. DeWitt, "Does Supreme Court Precedent Sink Submarine Patents?," *IDEA: The Journal of Law and Technology* 38 (1998): 601; Steve Blount, "The Use of Delaying Tactics to Obtain Submarine Patents and Amend Around a Patent That a Competitor Has Designed Around," *Journal of the Patent and Trademark Office Society* 81 (1999): 13–14.

52. Woodbury Patent Planing-Machine Company v. Keith, 482–484 (1879).
53. Robert C. Post, *Physics, Patents and Politics: A Biography of Charles Grafton Page* (New York: Science History Publications, 1976), 173–180.
54. William Greenleaf, *Monopoly on Wheels: Henry Ford and the Selden Automobile Patent* (Detroit: Wayne State University Press, 1961).
55. Lemelson received his patents between 1978 and 1994, beginning a worldwide campaign of litigation in 1989. The Lemelson patents were struck down in January 2004. Hiawatha Bray, "Cognex Wins Long-Running Patent Suit," *Boston Globe,* January 27, 2004, F3.
56. "A Strange Delay in the Patent Office," *New York Times,* March 28, 1888, 4; Kevin G. Rivette and David Kline, *Rembrandts in the Attic: Unlocking the Hidden Value of Patents* (Boston: Harvard Business School Press, 2000).
57. *United States of America v. American Bell Telephone Company and Emile Berliner. Brief for Appellees* (no printing details given: 1896), 112–116.
58. Drawbaugh's patent application had already been rejected once by the Patent Office, but his attorneys refused to treat the ruling as final.
59. *United States v. American Bell and Emile Berliner. Brief for Appellees,* 120–143; *United States of America v. American Bell Telephone Company and Emile Berliner. Brief for Appellants* (Boston: Addison C. Getchell, 1896), 48–61; American Bell Telephone Company and Emile Berliner v. United States, 68 F. 542, 558 (1st Cir. 1895).
60. *United States v. American Bell and Emile Berliner. Brief for Appellees,* 145–147, 151–155; *United States v. American Bell and Emile Berliner. Brief for Appellants,* 85–90.
61. George David Smith, *The Anatomy of a Business Strategy: Bell, Western Electric and the Origins of the American Telephone Industry* (Baltimore: Johns Hopkins University Press, 1985), 111–113; Report of Commissioner of Patents William Simonds on Kellogg's "condensed brief," December 13, 1892, RG60.3.2, year file 1892 / 11437, box 659, NARA.
62. Solicitor General Charles H. Aldrich to Attorney General William Miller, January 31, 1893, RG60.3.2, year file 1892 / 11437, file 1381, NARA.
63. Robert S. Taylor to Attorney General Judson Harmon, October 1, 1896, RG60.3.2, year file 1892 / 11437, file 15264, NARA.
64. Solicitor General Charles H. Aldrich to Attorney General William Miller, January 31, 1893, RG60.3.2, year file 1892 / 11437, file 1381, NARA.
65. Cummings and McFarland, *Federal Justice,* 308–309.
66. Robert S. Taylor to Attorney General William Miller, January 26, 1893, RG60.3.2, year file 1892 / 11437, file 1062, NARA.
67. U.S. Patent Office, *Annual Report of the Commissioner of Patents for the Year 1887* (Washington, DC: Government Printing Office, 1888); "Defects in the Patent Laws," *New York Times,* March 30, 1888, 2.
68. American Bell Telephone Company v. Dolbear, 15 F. 448, 453 (C.C.D. Mass. 1883).
69. American Bell Telephone Company, *Circuit Court of the United States for the First Circuit. United States of America v. American Bell Telephone Company and Emile Berliner. Brief for Defendants* (Boston: Alfred Mudge, 1894), 18.
70. United States v. American Bell Telephone Company and Emile Berliner, 65 F. 86, 88 (C.C.D. Mass. 1894).

71. Ibid., 91.
72. Robert S. Taylor to Attorney General Richard Olney, May 1, 1895, RG60.3.2, year file 1892 / 11437, file 6707, NARA.
73. American Bell Telephone Company and Emile Berliner v. United States, 546.
74. United States v. American Bell Telephone Company and Emile Berliner, 167 U.S. 224, 250 (1897). Justice Harlan dissented from the decision without writing an opinion; Justice Gray recused himself again, and Justice Wood also took no part in the case.
75. Ibid.
76. Herbert Hovenkamp, Mark D. Janis, and Mark A. Lemley, *I.P. and Antitrust: An Analysis of Antitrust Principles Applied to Intellectual Property Law* (New York: Aspen Law & Business, 2003), §1.3c, n. 11.
77. Henry v. A. B. Dick Company, 224 U.S. 1 (1912).
78. Victor Talking Machine Company v. The Fair, 123 F. 424, 426 (7th Cir. 1903).
79. See, e.g., Continental Paper Bag Company v. Eastern Paper Bag Company, 210 U.S. 405, 424–425 (1908); United States v. United Shoe Machinery Company of New Jersey, 247 U.S. 32, 57 (1918); Hartford-Empire Company v. United States, 323 U.S. 386, 433 (1945).
80. See James McNaught, counsel for Standard Telephone Company of Madison, Wisconsin, to Attorney General Judson Harmon, September 4, 1896, RG60.3.2, year file 1892 / 11437, file 14504, NARA; Causten Browne to James McNaught, September 16, 1896, RG60.3.2, year file 1892 / 11437, file 14602, NARA; "Independent Telephone Companies Organizing," *Western Electrician*, May 29, 1897, 306.
81. See, e.g., "The Effect of the Berliner Decision," *Electrical World*, May 15, 1897, 611–613; "Communications," *Western Electrician*, May 22, 1897, 291. Other electrical authorities concurred in this view: American Electrical Engineering Association, *Patented Telephony: A Review of the Patents Pertaining to Telephones and Telephonic Apparatus* (Chicago: 1897), 30.
82. Sherman Hoar to Attorney General, undated, RG60.3.2, year file 1892 / 11437, file 2270, NARA. The case in question was Miller v. Eagle Manufacturing Company, 151 U.S. 186 (1894).
83. American Bell Telephone Company v. National Telephone Manufacturing Company, 109 F. 976 (C.C.D. Mass. 1901); American Bell Telephone Company v. National Telephone Manufacturing Company, 119 F. 893 (1st Cir. 1903).
84. Stephen Skowronek, *Building a New American State: The Expansion of National Administrative Capacities, 1877–1920* (Cambridge: Cambridge University Press, 1982).
85. Usselman, *Regulating Railroad Innovation*, chap. 1; Richard R. John, "Farewell to the 'Party Period': Political Economy in Nineteenth-Century America," *Journal of Policy History* 16, no. 2 (2004): 117–125; Richard R. John, "Ruling Passions: Political Economy in Nineteenth-Century America," *Journal of Policy History* 18, no. 1 (2006).
86. Albert H. Walker, *Text-Book of the Patent Laws of the United States of America*, 2nd ed. (New York: L. K. Strouse, 1889), 135–136; H. E. Weisberger, "State Control over Patent Rights and Patented Articles," *Journal of the Patent Office Society* 20 (1938): 246–248.

87. Bement v. National Harrow Company, 186 U.S. 70, 91 (1902); A. Andrew Hauk, "Antitrust, Patents and Industrial Control: A Concretized Study in Public Regulation of Business" (JSD diss., Yale University School of Law, 1942), 116–142; Floyd L. Vaughan, *The United States Patent System: Legal and Economic Conflicts in American Patent History* (Norman: University of Oklahoma Press, 1956), 39–43.
88. Vaughan, *United States Patent System*, 46–49, 177–182.

5. Atlantic Crossings

1. U.S. Patent Office, *Annual Report of the Commissioner of Patents for the Year 1877* (Washington, DC: Government Printing Office, 1878).
2. The 1865 Patent Law Commission found that foreigners were responsible for approximately one-fifth of all applications—not issued patents—in 1852–1863. Stephen Van Dulken, *British Patents of Invention, 1617–1977: A Guide for Researchers* (London: The British Library, 1999), 87–88. The 1867–1869 data come from the *Report of the Select Committee on Letters Patent*, PP 1871 (368) X 603, 203. The 1884 data come from the *Annual Report of the Commissioner of Patents for 1887*, 5.
3. Robert H. Rines, "Some Areas of Basic Difference between United States Patent Law and That of the Rest of the World—and Why," *IDEA: The Journal of Law and Technology* 28, no. 1 (1987): 5–7.
4. Affidavit of J. Gordon Brown, undated, included in folder "Letters from James J. Storrow to Edward B. Brown, from May 12, 1887, to November 10, 1888," box 272, Bell Papers.
5. Robert V. Bruce, *Bell: Alexander Graham Bell and the Conquest of Solitude* (London: Gollancz, 1973), 162–166; A. Edward Evenson, *The Telephone Patent Conspiracy of 1876: The Elisha Gray-Alexander Bell Controversy and Its Many Players* (Jefferson, NC: McFarland, 2000), 59–64.
6. British Patent 4,765 of 1876, issued to William Morgan-Brown for "Electric telephony." The numbering of British patents began again at the beginning of each year, rather than forming a continuous numbered series as in the United States. As a communication from abroad, the first British telephone patent was issued in Morgan Brown's name rather than in Bell's.
7. Martin Daunton, *Royal Mail: The Post Office Since 1840* (London: Athlone Press, 1985), 319.
8. "The Quadruplex System in England," *Telegraphic Journal and Electrical Review*, September 1, 1878, 363.
9. E. C. Baker, *Sir William Preece, F.R.S.: Victorian Engineer Extraordinary* (London: Hutchinson, 1976), 162–171, 176.
10. Chief Engineer to Secretary of the Post Office, October 2, 1877, POST 30 / 330 / 2, BT Archives.
11. William Preece to Edward Graves, September 19, 1877, and memorandum by Graves, September 19, 1877, POST 30 / 330 / 3, BT Archives.
12. "The Inventor and the Official," *Telegraphic Journal and Electrical Review*, February 15, 1879, 68–69.
13. Memorandum by Graves, September 19, 1877, BT Archives, POST 30 / 330 / 3.

14. Chief Engineer to Secretary of the Post Office, December 5, 1877, BT Archives, POST 30 / 330 / 2; "Proposed arrangement with Colonel Reynolds for supply. Treasury withhold sanction pending expression of opinion by Law Officers," February–March 1878, BT Archives, POST 30 / 330 / 9.
15. F. G. C. Baldwin, *The History of the Telephone in the United Kingdom* (London: Chapman & Hall, 1925), 14–15; Bruce, *Bell*, 241–242.
16. Sir William Thomson to Alexander Graham Bell, August 30, 1877, box 128, Bell Papers.
17. Bruce, *Bell*, 243–244.
18. This agreement stated that until the second payment the balance of ownership should stand at 55 percent for Bell and Reynolds, 45 percent for the investors. It seems that the second installment may not have been paid, since in October 1879 the patentees' stake stood at 55 percent. Alexander Graham Bell to Gardiner Greene Hubbard, July 28, 1880, box 79, Bell Papers.
19. F. Warner to Alexander Melville Bell, March 23, box 5, 1878, Bell Papers.
20. "Summary of Capital and Shares in the Telephone Company Limited, 23 December 1878," BT 31 / 2433 / 12331, UKNA. Some details of the directorate in 1879 appear in Gardiner Greene Hubbard to Alexander Graham Bell and Mabel Hubbard Bell, August 11, 1879, box 79, Bell Papers and Gardiner Greene Hubbard to Alexander Graham Bell and Mabel Hubbard Bell, October 28, 1879, box 79, Bell Papers.
21. Charles A. Jones, "Great Capitalists and the Direction of British Overseas Investment in the Late Nineteenth Century: The Case of Argentina," *Business History* 22, no. 2 (1980): 154–156.
22. Gardiner Greene Hubbard to Alexander Graham Bell and Mabel Hubbard Bell, October 28, 1879, box 79, Bell Papers.
23. Gardiner Greene Hubbard to Alexander Graham Bell and Mabel Hubbard Bell, August 11, 1879, box 79, Bell Papers.
24. Alexander Graham Bell to the Directors of the Telephone Company, Limited, July 25, 1878, box 271, Bell Papers.
25. Ibid.
26. Gardiner Greene Hubbard to Alexander Graham Bell and Mabel Hubbard Bell, August 11, 1879, box 79, Bell Papers; Gardiner Greene Hubbard to Alexander Graham Bell and Mabel Hubbard Bell, October 28, 1879, box 79, Bell Papers; Alexander Graham Bell to the Directors of the Telephone Company, Limited, October 15, 1878, quoted in Bruce, *Bell*, 245.
27. Gardiner Greene Hubbard to Alexander Graham Bell, April 18, 1879, box 79, Bell Papers.
28. Gardiner Greene Hubbard to Alexander Graham Bell and Mabel Hubbard Bell, August 11, 1879, box 79, Bell Papers.
29. Ibid.
30. Gardiner Greene Hubbard to Alexander Graham Bell and Mabel Hubbard Bell, October 28, 1879, box 79, Bell Papers.
31. George Bernard Shaw, *The Irrational Knot* (London: Archibald Constable, 1905), ix–x.

32. Agreement between Western Union and Thomas Edison, May 31, 1878, HM 78, Edison Papers.
33. Menlo Park was the location of Edison's laboratory in New Jersey. He gained the moniker "the Wizard of Menlo Park" in 1878 during publicity surrounding the invention of the phonograph. See Paul Israel, *Edison: A Life of Invention* (New York: Wiley, 1998), 147.
34. Israel, *Edison*, 148–149, 185.
35. Thomas Edison to George Gouraud, June 24, 1878, LB003, Edison Papers.
36. George Gouraud to Thomas Edison, January 18, 1879, D7941, Edison Papers.
37. Arnold White to Thomas Edison, May 13, 1879, D7941, Edison Papers.
38. "Edison Telephone Company of London, Ltd., List of Stockholders," June 28, 1879, D7941, Edison Papers; George Gouraud to Thomas Edison, June 28, 1879, D7941, Edison Papers.
39. Edison Telephone Company of London, Ltd., Memorandum and Articles of Association, August 2, 1879, D7941, Edison Papers.
40. "Edison's Patents for Loud-Speaking Telephones. Agreement between Thomas Edison and Edward Pleydell Bouverie et al., July 14, 1879," D7941, Edison Papers.
41. George Gouraud to Thomas Edison, June 27, 1879, D7941, Edison Papers; Edward Johnson to Thomas Edison, July 20, 1879, D7941, Edison Papers.
42. "Agreement between Edison and Edison Telephone Company of London on Provincial Development," July 2, 1879, D7941, Edison Papers; George Gouraud to Thomas Edison, July 4, 1879, D7941, Edison Papers.
43. George Gouraud to Thomas Edison, July 8, 1879, D7941, Edison Papers.
44. "Edison's Patents for Loud-speaking Telephones. Agreement between Thomas Edison and Edward Pleydell Bouverie et al."
45. George Gouraud to Thomas Edison, July 14, 1879, Edison Papers, D7941, Edison Papers; Thomas Edison to George Gouraud, July 21, 1879, HM79, Edison Papers.
46. Edward Johnson to Charles Batchelor, October 21, 1879, D7941, Edison Papers.
47. Edward Johnson to Thomas Edison, December 28, 1879, D7941, Edison Papers; Edward Johnson to Thomas Edison, February 27, 1880, D8049, Edison Papers.
48. Baldwin, *History of the Telephone*, 32–34.
49. Gardiner Greene Hubbard to Alexander Graham Bell and Mabel Hubbard Bell, August 11, 1879, box 79, Bell Papers; Edward Johnson to Thomas Edison, September 30, 1879, D7941, Edison Papers.
50. George Gouraud to Thomas Edison, April 9 and 10, 1879, D7941, Edison Papers; Baldwin, *History of the Telephone*, 27, 33–35, 119.
51. Edward Johnson to Thomas Edison, March 22, 1880, D8049, Edison Papers.
52. Shaw, *Irrational Knot*, ix–x.
53. The progress of this case is detailed in "Attorney General v. Edison Telephone Company. Notes and Statements of Proceedings," POST 30 / 398, BT Archives.
54. Arnold White to Edward Johnson, December 21, 1879, D7941, Edison Papers.
55. Edward Johnson to Thomas Edison, December 28, 1879, D7941, Edison Papers.
56. Edward Johnson to Thomas Edison, November 17 and December 23, 1879, Edison Papers, D7941.

57. DeLancey Horton Louderback to Thomas Edison, December 18, 1879, D7941, Edison Papers.
58. "Mr Sanford or Hanford"—possibly the patent agent T. H. Handford. Thomas Edison to George Gouraud, December 24, 1879, D7941, Edison Papers.
59. Edison's marginalia on DeLancey Horton Louderback to Thomas Edison, December 18, 1879, D7941, Edison Papers.
60. Edward Johnson to Thomas Edison, February 19, 1880, D8049, Edison Papers.
61. Ibid.
62. Edward Johnson to Thomas Edison, February 19 and 27, 1880, D8049, Edison Papers.
63. "Edison Telephone Company—Memorandum," undated (February 1880), D8049, Edison Papers; Edward Johnson to Thomas Edison, April 5, 1880, D8049, Edison Papers.
64. Edward Johnson to Charles Batchelor, October 19, 1879, D7941, Edison Papers; Edward Johnson to Thomas Edison, February 21 and 27, March 4, 1880, D8049, Edison Papers.
65. Arnold White to Edward Johnson, April 1, 1880, D8049, Edison Papers; Edward Johnson to Thomas Edison, April 1, 1880, D8049, Edison Papers.
66. List of United Telephone Company shareholders, August 23, 1881, BT 31 / 2661 / 14163, UKNA.
67. See W. Bernard Carlson, "Entrepreneurship in the Early Development of the Telephone: How Did William Orton and Gardiner Hubbard Conceptualize This New Technology?," *Business and Economic History* 23, no. 2 (1994): 161–192.
68. For a review of the "electrical toy" theme in the American literature, see David Hochfelder, "Constructing an Industrial Divide: Western Union, AT&T, and the Federal Government, 1876–1971," *Business History Review* 76, no. 4 (2002): 708–710. On the attitude of the Post Office, see Charles R. Perry, *The Victorian Post Office: The Growth of a Bureaucracy* (Woodbridge, UK: Boydell Press for the Royal Historical Society, 1992).
69. Naomi R. Lamoreaux and Kenneth L. Sokoloff, "Inventors, Firms, and the Market for Technology in the Late Nineteenth and Early Twentieth Centuries," in *Learning by Doing in Markets, Firms, and Countries*, ed. Naomi R. Lamoreaux, Daniel M. G. Raff, and Peter Temin (Chicago: University of Chicago Press, 1999), 35–37, table 1.7.
70. See for example Nathan Rosenberg and Claudio Frischtak, eds., *International Technology Transfer: Concepts, Measures, and Comparisons* (New York: Praeger, 1985); David J. Jeremy, ed., *The Transfer of International Technology: Europe, Japan and the USA in the Twentieth Century* (Aldershot, UK: Edward Elgar, 1992).
71. For studies describing the international diffusion of telephone technology without reference to national patent situations, see James Foreman-Peck, "International Technology Transfer in Telephony, 1876–1914," in *International Technology Transfer: Europe, Japan and the USA, 1700–1914*, ed. David J. Jeremy (Aldershot, UK: Edward Elgar, 1991), 122–152; Helge Kragh, "Transatlantic Technology Transfer: The Reception and Early Use of the Telephone in the USA and Europe," in *European Historiography of Technology*, ed. Dan Ch. Christensen (Odense, Denmark: Odense University Press, 1993), 68–90.

72. Pierre Aulas, *Les Origines du Téléphone en France, 1876–1914* (Paris: Association pour le Développement de l'Histoire Économique, 1999); Robert D. MacDougall, *The People's Network: The Political Economy of the Telephone in the Gilded Age* (Philadelphia: University of Pennsylvania Press, 2013).

73. Wilfried Feldenkirchen, *Werner von Siemens: Inventor and International Entrepreneur* (Columbus: Ohio State University Press, 1994), 149–150; Werner Siemens to Alexander Graham Bell, November 29, 1877, quoted in Wolfgang Mache, "Reis-Telefon (1861/64) und Bell-Telefon (1875/77): Ein Vergleich," *Hessische Blätter für Volks- und Kulturforschung* 24 (1989): 49–50.

74. E. Feyerabend, *50 Jahre Fernsprecher in Deutschland 1877–1927* (Berlin: Reichspostministerium, 1927).

75. Artur Attman, Jan Kuuse, and Ulf Olsson, *L. M. Ericsson 100 Years*, vol. 1, *The Pioneering Years, Struggle for Concessions, Crisis, 1876–1932* (Stockholm: L. M. Ericsson, 1977), 47–109; Christian Jacobaeus, *L. M. Ericsson 100 Years*, vol. 3, *Evolution of the Technology, 1876–1976* (Stockholm: L. M. Ericsson, 1977), 417; Claes-Fredrik Helgesson, *Making a Natural Monopoly: The Configuration of a Techno-Economic Order in Swedish Telecommunications* (Stockholm: Stockholm School of Economics and the Economic Research Institute, 1999).

76. Herbert Laws Webb, *The Development of the Telephone in Europe* (London: Electrical Press, 1911); Foreman-Peck, "International Technology Transfer"; Kragh, "Transatlantic Technology Transfer." On the absence of patent protections in the Netherlands and Switzerland, see E. Schiff, *Industrialization without National Patents: The Netherlands, 1869–1912, Switzerland, 1850–1907* (Princeton, NJ: Princeton University Press, 1971).

77. Alexander Graham Bell to Gardiner Greene Hubbard, July 28, 1880, box 79, Bell Papers; Foreman-Peck, "International Technology Transfer," 133–134.

78. Gardiner Greene Hubbard to Alexander Graham Bell, January 8, 1881, box 79, Bell Papers; "The Oriental Telephone Company (Limited)," *Telegraphic Journal and Electrical Review*, January 7, 1882, 16.

6. Patent the Earth

1. J. E. Kingsbury, *The Telephone and Telephone Exchanges: Their Invention and Development* (London: Longmans, Green, 1915), 191, n. 2.

2. Ibid., 517.

3. On contemporary regard for Manzetti, see "Monument to 'The Inventor of the Telephone,'" *Telegraphic Journal and Electrical Review*, July 16, 1886, 68.

4. "Telephonic Litigation in America," *Telegraphic Journal and Electrical Review*, June 20, 1885, 558.

5. Hullet v. Hague, 2 B. & Ad. 370, 377 (1831), quoted in William M. Hindmarch, *A Treatise on the Law Relative to Patent Privileges for the Sole Use of Inventions* (London: Stevens, Norton and Benning, 1846), 120; Robert Frost, *A Treatise on the Law and Practice Relating to Letters Patent for Inventions* (London: Stevens & Haynes, 1891), 217–218, 395; H. I. Dutton, *The Patent System and Inventive Activity During the Industrial Revolution, 1750–1852* (Manchester: Manchester University Press, 1984), 77–80.

6. Charles Dickens, "A Poor Man's Tale of a Patent," *Household Words*, October 19, 1850, repr. in *The Works of Charles Dickens*, vol. 34 (New York: Scribner's, 1900), 113–119.
7. Adrian Johns, *Piracy: The Intellectual Property Wars from Gutenberg to Gates* (Chicago: University of Chicago Press, 2010), 260–261.
8. Klaus Boehm and Aubrey Silberston, *The British Patent System*, vol. 1: *Administration* (Cambridge: Cambridge University Press, 1967), 19–20, 29–30; Christine MacLeod, Jennifer Tann, James Andrew, and Jeremy Stein, "Evaluating Inventive Activity: The Cost of Nineteenth-Century UK Patents and the Fallibility of Renewal Data," *Economic History Review* 56, no. 3 (2003): 537–562.
9. John Hewish, *Rooms Near Chancery Lane: The Patent Office under the Commissioners, 1852–1883* (London: British Library, 2000).
10. Victor M. Batzel, "Legal Monopoly in Liberal England: The Patent Controversy in the Mid-Nineteenth Century," *Business History* 22, no. 2 (1980): 189–202; Moureen Coulter, *Property in Ideas: The Patent Question in Mid-Victorian Britain* (Kirksville, MO: Thomas Jefferson University Press, 1991); Johns, *Piracy*, 262–264.
11. Isambard Kingdom Brunel, "Memorandum for Evidence before the Select Committee of the House of Lords on the Patent Laws, 1851," quoted in Christine MacLeod, *Heroes of Invention: Technology, Liberalism and British Identity, 1750–1914* (Cambridge: Cambridge University Press, 2007), 265; Batzel, "Legal Monopoly," 191.
12. Johns, *Piracy*, 263–264.
13. Fritz Machlup and Edith Penrose, "The Patent Controversy in the Nineteenth Century," *Journal of Economic History* 10, no. 1 (1950): 4–5; Johns, *Piracy*, 267.
14. Johns, *Piracy*, 270–271.
15. MacLeod, *Heroes of Invention*, 251–264.
16. Ibid., 1–3, 91–131.
17. Christine MacLeod, "Concepts of Invention and the Patent Controversy in Victorian Britain," in *Technological Change: Methods and Themes in the History of Technology*, ed. Robert Fox (Amsterdam: Harwood Academic Press, 1996), 137–153.
18. Ibid., 141–147; Johns, *Piracy*, 272–273.
19. Johns, *Piracy*, 283.
20. *Annual Report of the Commissioner of Patents for 1906* (London: 1907).
21. W. R. Cornish, *Intellectual Property: Patents, Copyright, Trade Marks and Allied Rights* (London: Sweet & Maxwell, 1981), 83.
22. Brian Abel-Smith and Robert Stevens, *Lawyers and the Courts: A Sociological Study of the English Legal System 1750–1965* (Cambridge, MA: Harvard University Press, 1967), 51, 85–89; A. H. Manchester, *A Modern Legal History of England and Wales, 1750–1950* (London: Butterworths, 1980), 149–150.
23. House of Commons, Report from the Select Committee on Letters Patent, PP 1871 (368) X 603 (hereafter "Select Committee on Patents, 1871"), app. 3, 202.
24. Ibid., 97.
25. Frederick J. Bramwell, "The Society of Arts Patent Bill," *Journal of the Society of Arts*, September 30, 1881, p. 811.
26. *Reports of Patent Cases*, vols. 1–18 (London: Patent Office, 1884–1902).

27. Dirk Van Zyl Smit, "Professional Patent Agents and the Development of the English Patent System," *International Journal of the Sociology of Law* 13 (1985): 88–89.
28. Richard Webster (Lord Alverstone), *Recollections of Bar and Bench* (London: Edward Arnold, 1914), 186.
29. Hugh Fletcher Moulton, *The Life of Lord Moulton* (London: Nisbet, 1922); A. B. Schofield, *Dictionary of Legal Biography, 1845–1945* (Chichester, UK: Barry Rose, 1998), 48, 72; Nathan Wells, "Davey, Horace, Baron Davey (1833–1907)," *Oxford Dictionary of National Biography* (Oxford: Oxford University Press, 2004) (hereafter *ODNB*); T. M. Goodeve, *Abstract of Reported Cases Relating to Letters Patent for Inventions* (London: Henry Sweet, 1876), title page.
30. F. D. Mackinnon, "Webster, Richard Everard, Viscount Alverstone (1842–1915)," rev. N. G. Jones, *ODNB*.
31. "The Edison and Brush Appeal," *Telegraphic Journal and Electrical Review*, February 22, 1889, 198.
32. Thomas Terrell, *The Law and Practice relating to Letters Patent for Inventions*, 2nd ed. (London: Sweet & Maxwell, 1889), vii; Webster, *Recollections of Bar and Bench*, 186–187.
33. Moulton, *Life of Lord Moulton*, 46.
34. Terrell, *Law and Practice relating to Letters Patent*, vii; Stathis Arapostathis and Graeme Gooday, *Patently Contestable: Electrical Technologies and Inventor Identities on Trial in Britain* (Cambridge, MA: MIT University Press, 2013), 78.
35. Westinghouse v. Lancashire and Yorkshire Railway Company, 1 R. P. C. 98, 103 (1884).
36. "J. A. S.," "Patent Law," *Telegraphic Journal and Electrical Review*, September 24, 1886, 320; Henry Trueman Wood, "Sir Frederick Bramwell," in *Dictionary of National Biography*, 2nd supp. (New York: Macmillan, 1913), 213–216; Moulton, *Life of Lord Moulton*, 47; B. P. Cronin, "Bramwell, Sir Frederick Joseph, Baronet (1818–1903)," *ODNB*.
37. Arapostathis and Gooday, *Patently Contestable*, 69–71.
38. J. M. Rigg, "Jessel, Sir George (1824–1883)," *Dictionary of National Biography*, ed. Sidney Lee (New York: Macmillan, 1892), 29:368.
39. Terrell, *Law and Practice relating to Letters Patent*, vii; Iwan Rhys Morus, "Grove, Sir William Robert (1811–1896)," *ODNB*.
40. Alexander Graham Bell to Eliza Simmons Bell, October 25, 1876, box 27, Bell Papers.
41. British Patent 4,765 of 1876, issued to William Morgan-Brown for "Electric telephony," 13.
42. Alexander Graham Bell to Eliza Simmons Bell, October 25, 1876, box 27, Bell Papers.
43. "The British Association," *Engineering*, September 15, 1876, 242.
44. "Experiments in Telephony," *English Mechanic*, August 11, 1876, 551.
45. Morgan-Brown's Disclaimer and Memorandum of Alteration, February 13, 1878 (British Patent 4,765* of 1876).
46. British Patent 2,909 of 1877, issued to Thomas Edison for "Controlling by sound the transmission of electric currents and the reproduction of corresponding sounds at a distance."

47. British Patent 2,396 of 1878, issued to Thomas Edison for "Telephones and apparatus employed in electric circuits."
48. Edward Johnson to Thomas Edison, September 19, 1879, D7941, Edison Papers.
49. Edward Johnson to Thomas Edison, October 30, 1879, D7941, Edison Papers.
50. Disclaimer and Memorandum of Alteration of Thomas Edison, February 10, 1880 (British Patent 2,909* of 1877).
51. The Edison interests began, but did not complete, infringement proceedings over the Blake instrument, manufactured for the Telephone Company by the India Rubber, Gutta Percha and Telegraph Works Company. Edison Telephone Company of London v. India Rubber Co., L. R. 17 Ch. D. 137 (1881).
52. Disclaimer and Memorandum of Alteration of the United Telephone Company, June 13, 1881 (British Patent 2,909** of 1877).
53. Disclaimer and Memorandum of Alteration of the United Telephone Company, August 17, 1882 (British Patent 2,909*** of 1877). Justice Fry's opinion holding the Edison patent invalid is at United Telephone Company v. Harrison, Cox-Walker & Co., L. R. 21 Ch. D. 720, 745-747 (1882).
54. This point holds despite the fact that Edison's patent was later voided. Bell's problems of prior publication were well known and potentially fatal. The flaw in Edison's grant could be and was removed by disclaimer.
55. Edward Johnson to Thomas Edison, November 17, 1879, D7941, Edison Papers.
56. "Professor D. E. Hughes's Telephone, Microphone and Thermopile," *Engineer*, May 17, 1878, 343.
57. See, e.g., the letter signed "Right v. Might," *Telegraphic Journal and Electrical Review*, February 25, 1882, 143. For critical commentary on the proliferation of electrical patents and attendant litigation, see "The Infringement of Patent Rights," *Telegraphic Journal and Electrical Review*, November 1, 1879, 347-348; "Patents," *Telegraphic Journal and Electrical Review*, March 15, 1880, 95.
58. E. C. Baker, *Sir William Preece, F.R.S.: Victorian Engineer Extraordinary* (London: Hutchinson, 1976), 196.
59. F. G. C. Baldwin, *The History of the Telephone in the United Kingdom* (London: Chapman & Hall, 1925), 447.
60. Ibid., 102-103.
61. For the full record of the Maclean case, see *Telegraphic Journal and Electrical Review*, January 28, 1882, 65-70; February 4, 1882, 74-82; February 11, 1882, 98-106; February 18, 1882, 117-122; February 25, 1882, 136-139; March 4, 1882, 154-157. The Harrison, Cox-Walker case appears at ibid., May 6, 1882, 327-332; May 13, 1882, 350-352; May 20, 1882, 368-370; May 27, 1882, 377-386.
62. "United Telephone Company v. D. & G. Graham," *Telegraphic Journal and Electrical Review*, March 15, 1881, 103; "Agreement and Indenture between D. & G. Graham and David Graham Junior and the Provincial Telephone Company (Limited) and the National Telephone Company (Limited), May 1881," BT31 / 2765 / 15066, UKNA; "United Telephone Company v. Moseley & Sons," *Telegraphic Journal and Electrical Review*, August 1, 1881, 295; *Telegraphic Journal and Electrical Review*, November 5, 1881, 454.

63. "The United Telephone Company v. Alexander Maclean, Edinburgh," *Telegraphic Journal and Electrical Review*, January 7, 1882, 12. The first infringement suit under the Bell patent began in March 1878, but appears to have been settled after the denial of a motion for preliminary injunction. "The Bell Telephone," *Telegraphic Journal and Electrical Review*, March 15, 1878, 109.
64. "The United Telephone Company v. Alexander Maclean, Edinburgh," *Telegraphic Journal and Electrical Review*, January 7, 1882, 12; Letter Signed "Anti-Humbug," *Telegraphic Journal and Electrical Review*, January 21, 1882, 46-47.
65. "The Telephone Case," *Telegraphic Journal and Electrical Review*, January 28, 1882, 66.
66. "United Telephone Company v. Maclean," *Telegraphic Journal and Electrical Review*, February 11, 1882, 93.
67. Edward Johnson to Thomas Edison, July 20, 1879, D7941, Edison Papers.
68. Testimony of Sir Frederick Bramwell, "The Telephone Case," *Telegraphic Journal and Electrical Review*, January 28, 1882, 68; Opinion of Lord McLaren in ibid., March 4, 1882, 155.
69. "The Telephone Case," *Telegraphic Journal and Electrical Review*, March 4, 1882, 155.
70. Ibid., 155-156.
71. Argument of Mr. Mackintosh, "The Telephone Case," *Telegraphic Journal and Electrical Review*, February 18, 1882, 121.
72. See, e.g., "Recent Improvements in Telephones," *Engineering*, May 9, 1879, 387-388; J. Munro, "New Telephone Transmitters," *Electrician*, March 17, 1883, 425-428; Shelford Bidwell, "On Microphonic Contacts," *Journal of the Society of Telegraph Engineers and Electricians* 12, no. 48 (1883): 173-204, with discussion 208-240.
73. "The Telephone Case," *Telegraphic Journal and Electrical Review*, March 4, 1882, 155-156.
74. Ibid., 157.
75. Ibid.
76. "United Telephone Company v. Maclean," *Telegraphic Journal and Electrical Review*, February 11, 1882, 93.
77. *Telegraphic Journal and Electrical Review*, May 27, 1882, 386.
78. London and Globe Telephone and Maintenance Company, Ltd., Memorandum and Articles of Association; Additional Agreements to Purchase Patents in May, June and August 1882, BT31 / 2928 / 16324, UKNA; "City Notes—London and Globe Telephone Company," *Telegraphic Journal and Electrical Review*, July 7, 1883, 15-16.
79. On the London and Globe's acquisition of the Hunnings patent, see "Telephonic Litigation," *Telegraphic Journal and Electrical Review*, June 24, 1882, 464-465; Baldwin, *History of the Telephone*, 73-74.
80. Baldwin, *History of the Telephone*, 449-450; M. D. Fagen, ed., *A History of Engineering and Science in the Bell System: The Early Years, 1875-1925* (New York: Bell Telephone Laboratories, 1975), 74-78.
81. "The Telephone Case," *Telegraphic Journal and Electrical Review*, May 27, 1882, 382.

82. Ibid., 384.
83. Ibid., 384-385.
84. Baldwin, *History of the Telephone*, 72.
85. "The United Telephone Co. v. Harrison, Cox-Walker & Co.," *Telegraphic Journal and Electrical Review*, February 10, 1883, 115.
86. Arguments in the case appear in the *Telegraphic Journal and Electrical Review*, July 18, 1885, 48-56; July 25, 1885, 72-79; August 1, 1885, 97-102; August 8, 1885, 121-124; August 15, 1885, 146-150. Although Mr. Justice North rendered a verdict on August 12, 1885, he did not make the reasons for his judgment available until March 27, 1886; see *Telegraphic Journal and Electrical Review*, April 2, 1886, 302-305.
87. "United Telephone Company, Ltd. v. Bassano and Slater," *Telegraphic Journal and Electrical Review*, July 2, 1886, 13-14.
88. J. Munro, "Some Thoughts about the Telephone," *Telegraphic Journal and Electrical Review*, October 3, 1885, 290-291.
89. *Times*, July 6, 1886, 7; July 7, 1886, 6; July 17, 1886, 7; July 22, 1886, 4; August 17, 1886, 10.
90. Arapostathis and Gooday, *Patently Contestable*, 185-191.
91. "The Edison and Brush Appeal," *Telegraphic Journal and Electrical Review*, February 22, 1889, 198-199.
92. *Times* editorial, July 7, 1886, 11.
93. "English Telephone Patents," *Telegraphic Journal and Electrical Review*, July 2, 1886, 1.
94. *Pall Mall Gazette*, March 30, 1886, 3.
95. Letter of S. P. Thompson, *Times*, August 17, 1886, 10.
96. *Telegraphic Journal and Electrical Review*, May 27, 1882, 383; "The United Telephone Company v. Harrison, Cox-Walker & Co.," *Telegraphic Journal and Electrical Review*, February 10, 1883, 114.
97. *Telegraphic Journal and Electrical Review*, May 27, 1882, 383; "The United Telephone Company v. Harrison, Cox-Walker & Co.," *Telegraphic Journal and Electrical Review*, February 10, 1883, 113.
98. "Important Telephone Patent Case," *Telegraphic Journal and Electrical Review*, August 15, 1885, 147.
99. Ibid.
100. Otto v. Linford, 46 L. T. 35, 39 (1881).
101. "What Is a Filament?," *Telegraphic Journal and Electrical Review*, February 11, 1887, 121. See also James Roberts, *The Grant and Validity of British Patents for Inventions* (London: John Murray, 1903), 62.
102. Proctor v. Bennis, 4 R. P. C. 333 (1887), esp. the opinion of Lord Bowen at 359; Robert Frost, *A Treatise on the Law and Practice Relating to Letters Patent for Inventions*, 2nd ed. (London: Stevens & Haynes, 1898), 490.
103. James Johnson, *The Patentee's Manual: A Treatise on the Law and Practice of Patents for Inventions*, 6th ed. (London: Longmans, Green, 1890), 241-247.
104. Opinion of Lord Chelmsford in Harrison v. Anderston Foundry Co., L. R. 1 App. Cas. 574, 580 (1876).

105. W. R. Cornish and G. de N. Clark, *Law and Society in England, 1750–1950* (London: Sweet & Maxwell, 1989), 282.
106. P. S. Atiyah, *The Rise and Fall of Freedom of Contract* (Oxford: Clarendon Press, 1979), 388–389, 660–662; Steve Hedley, "Words, Words, Words: Making Sense of Legal Judgments, 1875–1940," in *Law Reporting in Britain*, ed. Chantal Stebbings (London: Hambledon Press, 1995), 169–186.
107. Lister v. Norton, 3 R. P. C. 199, 203 (1886).
108. "Patent Law Amendment Bill," *Journal of the Society of Arts*, December 9, 1881, 78.
109. Clark v. Adie, L. R. 2 App. Cas. 315, 320 (1873).
110. MacLeod, "Concepts of Invention," 140–141.
111. H. G. Fox, *Monopolies and Patents: A Study of the History and Future of the Patent Monopoly* (Toronto: University of Toronto Press, 1947), 238.
112. Clare Pettitt, *Patent Inventions: Intellectual Property and the Victorian Novel* (Oxford: Oxford University Press, 2004), 31–33; MacLeod, *Heroes of Invention*.
113. "The Edison and Brush Appeal," *Telegraphic Journal and Electrical Review*, February 22, 1889, 198.
114. Edward Manson, *The Builders of Our Law during the Reign of Queen Victoria* (London: Horace Cox, 1895), 161.
115. "The Telephone Case," *Telegraphic Journal and Electrical Review*, April 2, 1886, 302–303.
116. Letter of S. P. Thompson, *Times*, July 6, 1886, 7.
117. Select Committee on Patents, 1871, 16.
118. Speech of Mr. Samuelson, *Parliamentary Debates*, 3rd series, vol. 278, col. 370, April 16, 1883.
119. Letter of Prof. George Forbes, *Times*, July 22, 1886, 4. See also James Swinburne, "The Edison Filament Case," *Telegraphic Journal and Electrical Review*, August 6, 1886, 129.

7. Patents, Firms, and Systems

1. The Telephone Cases, 126 U.S. 1, 276 (1888).
2. Tim Wu, *The Master Switch: The Rise and Fall of Information Empires* (New York: Vintage, 2011).
3. Jonas Warren Stehman, *The Financial History of the American Telephone and Telegraph Company* (Boston: Houghton Mifflin, 1925), 51.
4. Richard R. John, *Network Nation: Inventing American Telecommunications* (Cambridge, MA: Belknap Press of Harvard University Press, 2010), 272, 386–395; Robert D. MacDougall, *The People's Network: The Political Economy of the Telephone in the Gilded Age* (Philadelphia: University of Pennsylvania Press, 2013), 231–245.
5. Alexander Graham Bell to "The Capitalists of the Electric Telephone Company," March 25, 1878, repr. in J. E. Kingsbury, *The Telephone and Telephone Exchanges: Their Invention and Development* (London: Longmans, Green, 1915), 89–92.
6. James D. Reid, *The Telegraph in America: Its Founders, Promoters and Noted Men* (New York: Derby Brothers, 1879), 634–636.

7. Kingsbury, *Telephone and Telephone Exchanges*, 90.
8. Thomas P. Hughes, *Networks of Power: Electrification in Western Society, 1880–1930* (Baltimore: Johns Hopkins University Press, 1983); Renate Mayntz and Thomas P. Hughes, eds., *The Development of Large Technical Systems* (Boulder, CO: Westview Press, 1988).
9. Alfred D. Chandler Jr. and James W. Cortada, "The Information Age: Continuities and Differences," in *A Nation Transformed by Information: How Information Has Shaped the United States from Colonial Times to the Present*, ed. Alfred D. Chandler Jr. and James W. Cortada (Oxford: Oxford University Press, 2000), 288.
10. Milton Mueller, *Universal Service: Competition, Interconnection, and Monopoly in the Making of the American Telephone System* (Cambridge, MA: MIT Press, 1997), 12–20.
11. Michael L. Katz and Carl Shapiro, "Network Externalities, Competition, and Compatibility," *American Economic Review* 75, no. 3 (1985): 424–440.
12. Alan Stone, *Public Service Liberalism: Telecommunications and Transitions in Public Policy* (Princeton, NJ: Princeton University Press, 1991), 126.
13. James Foreman-Peck and Robert Millward, *Public and Private Ownership of British Industry 1820–1990* (Oxford: Clarendon Press, 1994); Kenneth Lipartito, "'Cutthroat' Competition, Corporate Strategy, and the Growth of Network Industries," *Research on Technological Innovation, Management and Policy* 6 (1997): 1–53; Charles D. Jacobson, *Ties That Bind: Economic and Political Dilemmas of Urban Utility Networks, 1800–1990* (Pittsburgh, PA: University of Pittsburgh Press, 2000).
14. On the development of both theory and regulatory practice, see William W. Sharkey, *The Theory of Natural Monopoly* (Cambridge: Cambridge University Press, 1982); Thomas K. McCraw, *Prophets of Regulation: Charles Francis Adams, Louis D. Brandeis, James M. Landis, Alfred E. Kahn* (Cambridge, MA: Belknap Press of Harvard University Press, 1984); Thomas Hazlett, "The Curious Evolution of Natural Monopoly Theory," in *Unnatural Monopolies: The Case for Deregulating Public Utilities*, ed. R. W. Poole (Lexington, MA: Lexington Books, 1985); Herbert Hovenkamp, *Enterprise and American Law 1836–1937* (Cambridge, MA: Harvard University Press, 1991).
15. Foreman-Peck and Millward, *Public and Private Ownership;* Daniel T. Rodgers, *Atlantic Crossings: Social Politics in a Progressive Age* (Cambridge, MA: Belknap Press of Harvard University Press, 1998), chap. 4; Barbara Fried, *The Progressive Assault on Laissez Faire: Robert Hale and the First Law and Economics Movement* (Cambridge, MA: Harvard University Press, 1998), chap. 5.
16. Hughes, *Networks of Power,* chap. 2.
17. Kenneth Lipartito, "Culture and the Practice of Business History," *Business and Economic History* 24, no. 2 (1995): 26.
18. See, e.g., Steven W. Usselman, *Regulating Railroad Innovation: Business, Technology, and Politics in America, 1840–1920* (Cambridge: Cambridge University Press, 2002), pt. 2; Robert D. MacDougall, "The People's Telephone: The Politics of Telephony in the United States and Canada, 1876–1926" (PhD diss., Harvard University, 2004), 19–29.

19. Stehman, *Financial History*; J. H. Robertson, *The Story of the Telephone: A History of the Telecommunications Industry of Britain* (London: Isaac Pitman and Sons, 1947), 84.
20. James P. Baughman, "Written testimony on Bell System history," in Selected Testimony, United States v. American Telephone & Telegraph Co. et al. (552 F. Supp. 137: civil action no. 74-1698 [D.D.C.], 1982), vol. 1; Alfred E. Kahn, *The Economics of Regulation: Principles and Institutions* (New York: Wiley, 1971), 2:127.
21. Robert Bornholz and David S. Evans, "The Early History of Competition in the Telephone Industry," in *Breaking Up Bell: Essays on Industrial Organization and Regulation*, ed. David S. Evans (New York: North-Holland, 1983), 7-39.
22. Wiebe Bijker, Thomas P. Hughes, and Trevor J. Pinch, eds., *The Social Construction of Technological Systems* (Cambridge, MA: MIT Press, 1987); Merritt Roe Smith and Leo Marx, *Does Technology Drive History? The Dilemma of Technological Determinism* (Cambridge, MA: MIT Press, 1994); Trevor J. Pinch, "The Social Construction of Technology: A Review," in *Technological Change: Methods and Themes in the History of Technology*, ed. Robert Fox (Australia: Harwood Academic, 1996), 17-35.
23. Gerald Berk, *Alternative Tracks: The Constitution of American Industrial Order, 1865-1917* (Baltimore: Johns Hopkins University Press, 1994); Frank Dobbin, *Forging Industrial Policy: The United States, Britain, and France in the Railway Age* (Cambridge: Cambridge University Press, 1994); Richard R. John, "Elaborations, Revisions, Dissents: Alfred D. Chandler, Jr.'s *The Visible Hand* after Twenty Years," *Business History Review* 71, no. 2 (1997): 151-200; Lipartito, "Culture and the Practice of Business History"; Naomi R. Lamoreaux, Daniel M. G. Raff, and Peter Temin, "Beyond Markets and Hierarchies: Toward a New Synthesis of American Business History," *American Historical Review* 108, no. 2 (2003): 404-433.
24. Hughes, *Networks of Power*; Thomas P. Hughes, "The Evolution of Large Technical Systems," in Bijker, Hughes, and Pinch, *Social Construction of Technological Systems*, 51-82; Mayntz and Hughes, *Development of Large Technical Systems*; Berk, *Alternative Tracks*; Dobbin, *Forging Industrial Policy*.
25. Lipartito, "'Cutthroat' competition"; Mueller, *Universal Service*.
26. Lipartito, "Culture and the Practice of Business History"; Dan Schiller, "Social Movement in Telecommunications: Rethinking the Public Service History of U.S. Telecommunications, 1894-1919," *Telecommunications Policy* 22, nos. 4-5 (1998): 397-408; Richard R. John, "Theodore N. Vail and the Civic Origins of Universal Service," *Business and Economic History* 28, no. 2 (1999): 71-81; John, *Network Nation*; MacDougall, *People's Network*.
27. Kenneth Lipartito, "System Building at the Margin: The Problem of Public Choice in the Telephone Industry," *Journal of Economic History* 49, no. 2 (1989): 323-336; Lipartito, "'Cutthroat' Competition"; MacDougall, *People's Network*, 93.
28. Timothy W. Guinnane, William A. Sundstrom, and Warren Whatley, eds., *History Matters: Essays on Economic Growth, Technology, and Demographic Change* (Stanford, CA: Stanford University Press, 2004).
29. Andrew B. Jack, "The Channels of Distribution for an Innovation: The Sewing-Machine Industry in America, 1860-1865," *Explorations in Entrepreneurial History*

9 (1957): 113-141; Alfred D. Chandler Jr., *The Visible Hand: The Managerial Revolution in American Business* (Cambridge, MA: Belknap Press of Harvard University Press, 1977), 302-306.
30. Chandler, *Visible Hand,* 302-314.
31. Ibid., 374-375.
32. On exchange costs, see Kenneth Lipartito, *The Bell System and Regional Business: The Telephone in the South, 1877-1920* (Baltimore: Johns Hopkins University Press, 1989), 33-36, 39.
33. See, e.g., "National Telephone Company," *Electrician,* July 15, 1892; Lipartito, *Bell System and Regional Business,* 3.
34. John, *Network Nation,* 217-226; MacDougall, *People's Network,* 25-35.
35. W. Bernard Carlson, *Innovation as a Social Process: Elihu Thomson and the Rise of General Electric, 1870-1900* (Cambridge: Cambridge University Press, 1991), 9.
36. Robert W. Garnet, *The Telephone Enterprise: The Evolution of the Bell System's Horizontal Structure, 1876-1909* (Baltimore: Johns Hopkins University Press, 1985), 61.
37. Rosario J. Tosiello, *The Birth and Early Years of the Bell Telephone System, 1876-1880* (New York: Arno Press, 1979), 363-366; Garnet, *Telephone Enterprise,* 34-35; George David Smith, *The Anatomy of a Business Strategy: Bell, Western Electric and the Origins of the American Telephone Industry* (Baltimore: Johns Hopkins University Press, 1985), 44-45, 49-51.
38. Stehman, *Financial History,* 22; Garnet, *Telephone Enterprise,* 182, n. 49, quoting Henry L. Storke in 1909 testimony.
39. American Bell Telephone Company, *Annual Report for 1883* (Boston: 1884), 3.
40. Garnet, *Telephone Enterprise,* 62-66; Smith, *Anatomy of a Business Strategy,* 101-103. Few permanent licenses were issued before 1882.
41. American Bell, *Annual Report for 1881* (Boston: 1882), 2.
42. Dividends on the stock granted to Bell were usually suspended until the expiry dates of the original five-year contracts. Garnet, *Telephone Enterprise,* 70.
43. Federal Communications Commission (FCC), *Investigation of the Telephone Industry in the United States* (1939), 76th Cong., 1st Sess., H.Doc. 340, 149-150.
44. On the stock policies of electrical manufacturers, see Arthur A. Bright, *The Electric-Lamp Industry: Technological Change and Economic Development from 1800 to 1947* (New York: Macmillan, 1949), 94; Harold C. Passer, *The Electrical Manufacturers, 1875-1900: A Study in Competition, Entrepreneurship, Technical Change, and Economic Growth* (Cambridge, MA: Harvard University Press, 1953), 28-29, 69, 118-120, 155; Carlson, *Innovation as a Social Process,* 212-214.
45. Lipartito, *Bell System and Regional Business,* 55-60.
46. American Bell, *Annual Report for 1883,* 4.
47. Garnet, *Telephone Enterprise,* 68-69.
48. American Bell, *Annual Report for 1888* (Boston: 1889), 1; Garnet, *Telephone Enterprise,* 147.
49. FCC, *Investigation of the Telephone Industry,* 22.
50. American Bell, *Annual Reports* (Boston: 1883-1895); Stehman, *Financial History,* 39; Federal Communications Commission, *Proposed Report of the Telephone*

Investigation. Pursuant to Public Resolution No. 8, 74th Congress (Washington, DC: U.S. Government Printing Office, 1938), 55.
51. FCC, *Investigation of the Telephone Industry*, 19, 55, 60.
52. Garnet, *Telephone Enterprise*, chap. 5.
53. American Bell, *Annual Report for 1885* (Boston: 1886), 24.
54. Theodore Vail to W. H. Forbes, March 28, 1881, repr. as Appendix B in Federal Communications Commission, *Report on the Engineering and Research Departments of the Bell System*, vol. 1 (Telephone investigation, special investigation docket No. 1) (Washington, DC: 1937).
55. Garnet, *Telephone Enterprise*, 62–63.
56. Smith, *Anatomy of a Business Strategy*, 59–60, 86–88.
57. FCC, *Report on the Engineering and Research Departments*, 1:9–10.
58. Ibid., 1:14–16, 22–27.
59. Smith, *Anatomy of a Business Strategy*, 113–117.
60. Leonard S. Reich, *The Making of American Industrial Research: Science and Business at G.E. and Bell, 1876–1926* (Cambridge: Cambridge University Press, 1985), chaps. 7–8.
61. Garnet, *Telephone Enterprise*, 85; American Bell, *Annual Report for 1885*, 5.
62. "Bell Telephone Conference," *Electrical World*, May 23, 1885, and June 13, 1885.
63. American Bell, *Annual Report for 1885*, 5–6.
64. Ibid., 5.
65. Kingsbury, *Telephone and Telephone Exchanges*, 420–421; Garnet, *Telephone Enterprise*, 80–81; David Gabel, "Divestiture, Spin-Offs, and Technological Change in the Telecommunications Industry: A Property Rights Analysis," *Harvard Journal of Law and Technology* 3 (1990): 78–79.
66. John, *Network Nation*, 214; MacDougall, *People's Network*, 82–83.
67. American Bell, *Annual Report for 1892* (Boston: 1893), 5–6; David F. Weiman, "Building Universal Service in the Early Bell System: The Co-Evolution of Regional Urban Systems and Long Distance Telephone Networks," in Guinnane, Sundstrom, and Whatley, *History Matters*, 328–363.
68. Lipartito, *Bell System and Regional Business*, chap. 4.
69. Garnet, *Telephone Enterprise*, 89, 92–93; Milton Mueller, "The Switchboard Problem: Scale, Signaling, and Organization in Manual Telephone Switching, 1877–1897," *Technology and Culture* 30, no. 3 (1989): 543–544; Stephen B. Adams and Orville R. Butler, *Manufacturing the Future: A History of Western Electric* (Cambridge: Cambridge University Press, 1999), 57; John, *Network Nation*, 221; MacDougall, *People's Network*, 82.
70. American Bell, *Annual Report for 1893* (Boston: 1894), 14.
71. George David Smith, *The Anatomy of a Business Strategy: Bell, Western Electric and the Origins of the American Telephone Industry* (Baltimore: Johns Hopkins University Press, 1985), 73.
72. On the combination of patents with other forms of intellectual property and with other proprietary strategies, see Mira Wilkins, "The Neglected Intangible Asset: The Influence of the Trade Mark on the Rise of the Modern Corporation," *Busi-

ness History 34, no. 1 (1992): 66-95; Gideon Parchomovsky and Peter Siegelman, "Towards an Integrated Theory of Intellectual Property," *Virginia Law Review* 88, no. 7 (2002): 1455-1528.
73. Victor S. Clark, *History of Manufactures in the United States*, vol. 2, 1860-1893 (New York: McGraw-Hill, 1929), 381.
74. Ronald Coase, "The Nature of the Firm," *Economica* 4, no. 16 (1937): 386-405; Oliver Williamson, *Markets and Hierarchies, Analysis and Antitrust Implications: A Study in the Economics of Internal Organization* (New York: Free Press, 1975); Naomi R. Lamoreaux, Daniel M. G. Raff, and Peter Temin, "Beyond Markets and Hierarchies: Toward a New Synthesis of American Business History," *American Historical Review* 108, no. 2 (2003).

8. Patents and the Networked Nation

1. Many contemporary sources counted each subscriber station as two instruments or "telephones" (transmitter and receiver). I follow the modern usage and consider each station a single telephone.
2. American Bell Telephone Company, *Annual Reports* (Boston: 1881-1895).
3. American Bell, *Annual Report for 1893* (Boston: 1894), 10.
4. Robert Bornholz and David S. Evans, "The Early History of Competition in the Telephone Industry," in *Breaking Up Bell: Essays on Industrial Organization and Regulation*, ed. David S. Evans (New York: North-Holland, 1983), 25.
5. Jonas Warren Stehman, *The Financial History of the American Telephone and Telegraph Company* (Boston: Houghton Mifflin, 1925), 72; Richard Gabel, "The Early Competitive Era in Telephone Communication, 1893-1920," *Law and Contemporary Problems* 34, no. 2 (1969): 343.
6. Robert W. Garnet, *The Telephone Enterprise: The Evolution of the Bell System's Horizontal Structure, 1876-1909* (Baltimore: Johns Hopkins University Press, 1985), 88; David Gabel, "The Evolution of a Market: The Emergence of Regulation in the Telephone Industry of Wisconsin, 1893-1917" (PhD diss., University of Wisconsin, 1987), 51-52.
7. American Bell, *Annual Report for 1884* (Boston: 1885), 4; Gabel, "Evolution of a Market," chap. 4; Robert D. MacDougall, *The People's Network: The Political Economy of the Telephone in the Gilded Age* (Philadelphia: University of Pennsylvania Press, 2013), 83-84.
8. MacDougall, *People's Network*, 90.
9. American Bell, *Annual Report for 1893*, 10.
10. Remarks by Mr. Gifford of the Ohio Valley Telephone Company in Louisville, Kentucky, National Telephone Exchange Association (NTEA), *Eighth Annual Meeting of the National Telephone Exchange Association, September 7-10, 1886* (Brooklyn, NY: 1886), 127-128.
11. "The Crowning Achievements of the Telephone," *Scientific American*, February 18, 1893, 98.
12. "The Wires Must Be Buried," *New York Times*, September 14, 1886, 4; Frederick Leland Rhodes, "How the Telephone Wires Were First Put Underground," *Bell*

Telephone Quarterly 2 (1923): 240–254; Richard R. John, *Network Nation: Inventing American Telecommunications* (Cambridge, MA: Belknap Press of Harvard University Press, 2010), 222–226; MacDougall, *People's Network*, 31.

13. American Bell, *Annual Report for 1893*, 9.
14. Ibid., 11.
15. Milton Mueller, *Universal Service: Competition, Interconnection, and Monopoly in the Making of the American Telephone System* (Cambridge, MA: MIT Press, 1997), 57; MacDougall, *People's Network*, 29–30.
16. House of Commons, Report from the Select Committee on the Telephone Service, PP 1895 (350) XIII 21, app. 2, 300, figs. at January 1, 1895; MacDougall, *People's Network*, 90.
17. John, *Network Nation*, 238–268; MacDougall, *People's Network*, 95–101.
18. MacDougall, *People's Network*, 95; John, *Network Nation*, 241; Basilio Catania, "The U.S. Government versus Alexander Graham Bell: An Important Acknowledgment for Antonio Meucci," *Bulletin of Science, Technology and Society* 22, no. 6 (2002): 428.
19. Milton Mueller, "The Switchboard Problem: Scale, Signaling, and Organization in Manual Telephone Switching, 1877–1897," *Technology and Culture* 30, no. 3 (1989); Kenneth Lipartito, "When Women Were Switches: Technology, Work, and Gender in the Telephone Industry, 1890–1920," *American Historical Review* 99, no. 4 (1994): 1082.
20. "The Telephone," *Western Electrician*, October 15, 1887, 193; John, *Network Nation*, 241–244; MacDougall, *People's Network*, 96–98.
21. John, *Network Nation*, 250–253; MacDougall, *People's Network*, 37–41.
22. American Bell, *Annual Report for 1885* (Boston: 1886), 14.
23. The federal census of electrical industries in 1902 turned up eighty-nine non-Bell telephone systems claiming to have existed before 1894—including thirty-three claiming origins in the 1880s—but it is unclear whether these had operated continuously out of sight of the patentee or had been shut down by lawsuits and revived after the patent expired. U.S. Bureau of the Census, *Telephones and Telegraphs, 1902* (Washington, DC: U.S. Government Printing Office, 1906), tables 10 and 12.
24. NTEA, *Sixth Annual Meeting of the National Telephone Exchange Association, September 16–17, 1884* (Brooklyn, NY: 1884), 145.
25. Harry B. MacMeal, *The Story of Independent Telephony* (Chicago: Independent Pioneer Telephone Association, 1934), 26–33; Kenneth Lipartito, *The Bell System and Regional Business: The Telephone in the South, 1877–1920* (Baltimore: Johns Hopkins University Press, 1989), 83; MacDougall, *People's Network*, 40–41.
26. American Bell's leading patent expert at the time had "no doubt that Philadelphia has borne the brunt" of infringement nationwide: NTEA, *Sixth Annual Meeting*, 145.
27. Catania, "U.S. Government versus Alexander Graham Bell," 428; "The Telephone," *Western Electrician*, January 21, 1888, 35; John, *Network Nation*, 246, 254.
28. "The Awakening Telephone Business," *Electrical Review*, January 24, 1894, 42.
29. John Brooks, *Telephone: The First Hundred Years* (New York: Harper & Row, 1976), 99.

30. James J. Storrow to John E. Hudson, November 17, 1891, quoted in N. R. Danielian, *A.T.&T.: The Story of Industrial Conquest* (New York: Vanguard Press, 1939), 97.
31. "The Bell Company in the Patent Office," *New York Times*, April 18, 1891, 4.
32. "James J. Storrow," *Western Electrician*, April 24, 1897, 228.
33. "Important Decisions Expected," *New York Times*, March 4, 1895, 15.
34. "The Telephone Patent Situation," *Electrical Engineer*, January 17, 1894, 44–46; "Many Patents Are at Stake," *New York Times*, November 16, 1894, 7.
35. Charles H. Aldrich, *The American Bell Telephone Monopoly and the Pending Legislation in Its Interest: A Memorial to the Fifty-Third Congress* (Chicago: 1894), 3.
36. Bate Refrigerating Co. v. Sulzberger, 157 U.S. 1 (1895).
37. American Bell, *Annual Report for 1894* (Boston: 1895), 13.
38. Federal Communications Commission (FCC), *Investigation of the Telephone Industry in the United States* (1939), 76th Cong., 1st Sess., H.Doc. 340, 216.
39. Naomi R. Lamoreaux and Kenneth L. Sokoloff, "Inventors, Firms, and the Market for Technology in the Late Nineteenth and Early Twentieth Centuries," in *Learning by Doing in Markets, Firms, and Countries*, ed. Naomi R. Lamoreaux, Daniel M. G. Raff, and Peter Temin (Chicago: University of Chicago Press, 1999), 41–42.
40. Lamoreaux and Sokoloff, "Inventors, Firms, and the Market," 41 n. 27, 42.
41. John, *Network Nation*, 280–281.
42. Western Electric Co. v. Home Telephone Co., 85 F. 649 (C.C.S.D. Ala. 1898); Capital Telephone & Telegraph Co., 86 F. 769 (C.C.N.D. Cal. 1898).
43. FCC, *Investigation of the Telephone Industry*, 217.
44. Western Electric Co. v. Western Telephone Construction Co., 81 F. 572 (C.C.N.D. Ill. 1897); John, *Network Nation*, 316–317.
45. Editorial, *Western Electrician*, May 15, 1897, 274.
46. John, *Network Nation*, 315–316.
47. U.S. Bureau of the Census, *Telephones and Telegraphs, 1902*, table 10.
48. *Electrical World*, May 22, 1897, 651.
49. Correlli Barnett, *The Audit of War: The Illusion and Reality of Britain as a Great Nation* (London: Macmillan, 1986).
50. Grosvenor Lowrey, "The Telephone," *Electrical Review*, January 25, 1893, 7–8.
51. Joan Nix and David Gabel, "AT&T's Strategic Response to Competition: Why Not Preempt Entry?," *Journal of Economic History* 53, no. 2 (1993): 377–387.
52. U.S. Bureau of the Census, *Telephones: 1907* (Washington, DC: U.S. Government Printing Office, 1910), tables 8 and 9. The average number of telephones on each farmer line was just eleven. Ibid., 23.
53. Lipartito, *Bell System and Regional Business*, 108–109; MacDougall, *People's Network*, 47.
54. "The Story of Indiana," *Telephony*, February 1907, 106.
55. U.S. Bureau of the Census, *Telephones: 1907*, table 5.
56. MacMeal, *Story of Independent Telephony*, 26.
57. Mueller, *Universal Service*, 57–59.
58. MacDougall, *People's Network*, 112.
59. FCC, *Investigation of the Telephone Industry*, 133–135.

60. W. A. Jackson, President of the Central Union Telephone Company, to J. E. Hudson, President of American Bell, 1899, quoted in Mueller, *Universal Service*, 70, n. 43.
61. "New York Notes," *Western Electrician*, October 29, 1887, 216; Editorial, *Western Electrician*, May 24, 1890, 286.
62. "How New Yorkers Communicate with Each Other and with the World at Large," *Electrical Review*, October 22, 1892, 100.
63. Mueller, *Universal Service*, 67.
64. LeRoy W. Stanton, "How the Bell Lost Its Grip," in *Telephone Development: Scope and Effect of Competition*, ed. Vinton A. Sears (Boston: Barta Press, 1905), 103–106; MacDougall, *People's Network*, 145.
65. Mueller, *Universal Service*, 60–61.
66. The percentage of "duplicate" subscriptions in competitive cities was typically between 10 and 20 percent, although it reached almost 40 percent in at least one outlier (Saginaw, Michigan). New York Telephone Company, *Telephone Competition from the Standpoint of the Public* (New York: 1906), 14.
67. Kenneth Lipartito, "'Cutthroat' Competition, Corporate Strategy, and the Growth of Network Industries," *Research on Technological Innovation, Management and Policy* 6 (1997); Mueller, *Universal Service*.
68. Thomas Lockwood, "Ten Years of Progress in Practical Telephony," in *Ninth Annual Meeting of the National Telephone Exchange Association, September 26–27, 1887* (Brooklyn, NY: 1887), 53.
69. Ibid.
70. U.S. Bureau of the Census, *Telephones and Telegraphs, 1912* (Washington, DC: U.S. Government Printing Office, 1914), table 13.
71. U.S. Bureau of the Census, *Telephones and Telegraphs, 1912*, table 2.
72. Chicago City Council, *Telephone Service and Rates: Report of the Committee on Gas, Oil and Electric Light to the City Council of Chicago* (Chicago: 1907), 180–181, 191–192; U.S. Bureau of the Census, *Telephones and Telegraphs, 1912*, table 2.
73. John, *Network Nation*, 295–299; MacDougall, *People's Network*, 117–118; Claude S. Fischer, *America Calling: A Social History of the Telephone to 1940* (Berkeley, CA: University of California Press, 1992), 50.
74. Robert D. MacDougall, "The People's Telephone: The Politics of Telephony in the United States and Canada, 1876–1926" (PhD diss., Harvard University, 2004), 101.
75. The full story is told in John, *Network Nation*, chaps. 8–11; MacDougall, *People's Network*, 194–218.
76. Editorial, *New York Times*, February 18, 1890, 4; Stehman, *Financial History*, 80.
77. Chicago City Council and William J. Hagenah, *Report on the Investigation of the Chicago Telephone Company* (Chicago: 1911); Mueller, *Universal Service*, 56–62; David Gabel, "Competition in a Network Industry: The Telephone Industry, 1894–1910," *Journal of Economic History* 54, no. 3 (1994): 560–564; John, *Network Nation*, 323–327; MacDougall, *People's Network*, 201–202.
78. Gabel, "Competition in a Network Industry," 553; MacDougall, "People's Telephone," 198.
79. Quoted in Gabel, "Competition in a Network Industry," 550.

80. David F. Weiman and Richard C. Levin, "Preying for Monopoly? The Case of Southern Bell Telephone Company, 1894-1912," *Journal of Political Economy* 102, no. 1 (1994): 103-126; Gabel, "Competition in a Network Industry," 548.
81. FCC, *Investigation of the Telephone Industry*, 21.
82. Lipartito, *Bell System and Regional Business*, 105; Gabel, "Competition in a Network Industry," 548; Lipartito, "'Cutthroat' Competition," 27-29.
83. MacDougall, *People's Network*, 140.
84. "Our Rural Friends," *Telephony*, May 1907, 304-305.
85. FCC, *Investigation of the Telephone Industry*, 137-138; Lipartito, *Bell System and Regional Business*, 134-140.
86. "Hugh Dougherty, Traitor," *American Telephone Journal*, June 3, 1905, 358.
87. John, *Network Nation*, 318-322; MacDougall, *People's Network*, 199-205.
88. New York Telephone, *Telephone Competition*, title page.
89. Kenneth Lipartito, "System Building at the Margin: The Problem of Public Choice in the Telephone Industry," *Journal of Economic History* 49, no. 2 (1989); Mueller, *Universal Service*, chaps. 9-11.
90. MacDougall, *People's Network*, 195-199.
91. Alfred D. Chandler Jr., *Scale and Scope: The Dynamics of Industrial Capitalism* (Cambridge, MA: Belknap Press of Harvard University Press, 1990), 227-228.
92. Johann Peter Murmann, *Knowledge and Competitive Advantage: The Coevolution of Firms, Technology, and National Institutions* (Cambridge: Cambridge University Press, 2003).
93. Walton H. Hamilton, *Patents and Free Enterprise* (Washington, DC: U.S. Government Printing Office, 1941), 93.
94. Gary Jacobson and John Hillkirk, *Xerox: American Samurai* (New York: Macmillan, 1987), 75-80; Herbert A. Johnson, "The Wright Patent Wars and Early American Aviation," *Journal of Air Law and Commerce* 69, no. 1 (2004): 43-48.
95. Douglas J. Puffert, "Path Dependence, Network Form, and Technological Change," in *History Matters: Essays on Economic Growth, Technology, and Demographic Change*, ed. Timothy W. Guinnane, William A. Sundstrom, and Warren Whatley (Stanford, CA: Stanford University Press, 2004), 63-95.
96. Chandler, *Scale and Scope*, 227; Alfred D. Chandler Jr., *The Visible Hand: The Managerial Revolution in American Business* (Cambridge, MA: Belknap Press of Harvard University Press, 1977), 374-375.

Conclusion

1. C. F. Kettering, ed., *Centennial Celebration of the American Patent System* (Washington, DC: U.S. Government Printing Office, 1936), 58.
2. Ibid., 51-62.
3. Wolfgang Mache, "Reis-Telefon (1861/64) und Bell-Telefon (1875/77): Ein Vergleich," *Hessische Blätter für Volks- und Kulturforschung* 24 (1989): 45-62.
4. U.S. House of Representatives, H. Res. 269, 107th Congress (2002); House of Commons of Canada, *Journals*, 37th Parliament, 1st Sess., no. 211, June 21, 2002.

5. David A. Hounshell, "Elisha Gray and the Telephone: On the Disadvantages of Being an Expert," *Technology and Culture* 16, no. 2 (1975): 133–161; Thomas P. Hughes, *American Genesis: A Century of Invention and Technological Enthusiasm, 1870–1970* (New York: Viking, 1989), 15–16; Michael E. Gorman and W. Bernard Carlson, "Interpreting Invention as a Cognitive Process: The Case of Alexander Graham Bell, Thomas Edison, and the Telephone," *Science, Technology, & Human Values* 15, no. 2 (1990): 131–164; Michael E. Gorman, "Mind in the World: Cognition and Practice in the Invention of the Telephone," *Social Studies of Science* 27, no. 4 (1997): 583–624.
6. N. R. Danielian, *A.T.&T.: The Story of Industrial Conquest* (New York: Vanguard Press, 1939), 3–4.
7. Kettering, *Centennial Celebration*, 66.
8. Hughes, *American Genesis;* Alfred D. Chandler, *Shaping the Industrial Century: The Remarkable Story of the Evolution of the Modern Chemical and Pharmaceutical Industries* (Cambridge, MA: Harvard University Press, 2005).
9. Walton H. Hamilton, *Patents and Free Enterprise* (Washington, DC: U.S. Government Printing Office, 1941), 7.
10. Robert Lynd, quoted in David F. Noble, *America by Design: Science, Technology, and the Rise of Corporate Capitalism* (New York: Knopf, 1977), 109.
11. Kettering, *Centennial Celebration*, 6–7.
12. Lillian Hoddeson, "The Emergence of Basic Research in the Bell Telephone System, 1875–1915," *Technology and Culture* 22 (1981): 529–537; Leonard S. Reich, *The Making of American Industrial Research: Science and Business at G.E. and Bell, 1876–1926* (Cambridge: Cambridge University Press, 1985).
13. Hughes, *American Genesis*, 15.
14. See, e.g., Reese Jenkins, *Images and Enterprise: Technology and the American Photographic Industry, 1839 to 1925* (Baltimore: Johns Hopkins University Press, 1975), 184; Alfred D. Chandler, *The Visible Hand: The Managerial Revolution in American Business* (Cambridge, MA: Belknap Press of Harvard University Press, 1977), 374–375.
15. Ruth Brandon, *A Capitalist Romance: Singer and the Sewing Machine* (Philadelphia: J. B. Lippincott, 1977), 95–99.
16. Reich, *Making of American Industrial Research*, 53.
17. Federal Communications Commission (FCC), *Investigation of the Telephone Industry in the United States* (1939), 76th Cong., 1st Sess., H.Doc. 340, 213–214.
18. Danielian, *A.T.&T.*, 126–133; FCC, *Investigation of the Telephone Industry*, 224–231; Reich, *Making of American Industrial Research*, 235–238.
19. The number of patent suits filed each year doubled during the 1990s, and has since continued to rise sharply, from around 2,500 in the year 2000 to over 4,000 in 2011 and over 5,000 in 2012. http://www.uscourts.gov/Statistics/StatisticalTablesForThe FederalJudiciary/StatisticalTables_Archive.aspx.
20. Richard N. Langlois, "The Vanishing Hand: The Changing Dynamics of Industrial Capitalism," *Industrial and Corporate Change* 12, no. 2 (2003): 351–385; Naomi R. Lamoreaux, Daniel M. G. Raff, and Peter Temin, "Beyond Markets and Hierarchies: Toward a New Synthesis of American Business History," *American Historical*

Review 108, no. 2 (2003): 405; Eric Hintz, "The Post-Heroic Generation: American Independent Inventors, 1900–1950," *Enterprise and Society* 12, no. 4 (2011): 732–748; Naomi R. Lamoreaux, Kenneth L. Sokoloff, and Dhanoos Sutthiphisal, "Patent Alchemy: The Market for Technology in U.S. History," *Business History Review* 87 (2013): 34–38.

21. Samuel Kortum and Josh Lerner, "Stronger Protection or Technological Revolution: What Is Behind the Recent Surge in Patenting?," NBER Working Paper 6204 (1997); Adam B. Jaffe and Josh Lerner, *Innovation and Its Discontents: How Our Broken Patent System Is Endangering Innovation and Progress, and What to Do about It* (Princeton, NJ: Princeton University Press, 2004).
22. Jaffe and Lerner, *Innovation and Its Discontents*, 56.
23. Ibid., 19.

Acknowledgments

Like a patent, a book draws on the ideas and support of many people whose names do not—as a peculiar requirement of the form—appear under the title. With that in mind I offer heartfelt thanks to those friends, colleagues, and members of my family who have helped see this project through.

Students, mentors, and friends at many institutions have supported this work and its author. Martin Daunton at Cambridge University guided my earliest research on the telephone with great generosity and broad learning. Bill Nelson at NYU Law School and Sally Gordon at the University of Pennsylvania Law School each gave me an academic home and allowed me to run up debts, intellectual and personal, that I surely cannot repay. A year at Princeton's Program in Law and Public Affairs, talking and teaching with Dirk Hartog, was profoundly formative. My colleagues and students at Brooklyn Law School have provided an intellectually dynamic and wonderfully collegial setting in which to finish this book. Among the others who have shared their thoughts and insights along the way, I am particularly grateful to Lionel Bently, Richard Bernstein, Fred Bloom, Oren Bracha, Bill Cornish, Rochelle Dreyfuss, David Edgerton, Catherine Fisk, Lawrence Friedman, David Gabel, Graeme Gooday, Robert Gordon, David Hochfelder, Daniel Hulsebosch, Paul Israel, Herb Johnson, Zorina Khan, Naomi Lamoreaux, Sophia Lee, Kenneth Lipartito, Christine MacLeod, Peter Martland, Nick Parrillo, Pedro Ramos

Pinto, Gautham Rao, Kim Scheppele, Chris Serkin, Jonathan Silberstein-Loeb, Katherine Strandburg, Karen Tani, Adam Tooze, Steven Usselman, Polk Wagner, and Tess Wilkinson-Ryan. Special thanks are due to Robert MacDougall and Richard John, two scholars who showed me how the history of the telephone in society can and should be written.

For guidance on research and useful sources, I am indebted to Ray Martin and the staff of the British Telecom archives, and to Steven van Dulken, Neil Johannessen, and Bill Caughlin. I am especially grateful to Sheldon Hochheiser, who provided expert advice and assistance during a time of considerable upheaval at the AT&T archives.

Funding for research and writing was provided by the Arts and Humanities Research Board; by Trinity Hall, Cambridge; and by an Anniversary Fellowship from the Economic History Society. I have received additional funding for my work on patent history from the William Nelson Cromwell Foundation, from Princeton University's Program in Law and Public Affairs, and from Brooklyn Law School. Their support is much appreciated.

At Harvard University Press, the editorial hand of Joyce Seltzer and the assistance of Brian Distelberg have steered this book to completion with great patience and care. Joyce's comments improved the book in innumerable ways, as did the thoughtful comments of two anonymous readers for the press. Much of Chapter 3 previously appeared as "Who Invented the Telephone? Lawyers, Patents, and the Judgments of History," *Technology & Culture* 51, no. 4 (2010): 854–878 (copyright © 2010 The Society for the History of Technology, reprinted with permission by Johns Hopkins University Press). In that form, it benefited from the insights of the journal's editor, John Staudenmaier, and four anonymous referees.

My other thanks are owed for reasons that touch upon this book, but go far beyond it. I will forever be grateful for the constant support of my family, and especially of my parents, Sue and Tony Beauchamp. Their help and encouragement over the years have been more meaningful to me than they know.

"I am sick of the Telephone and sick of Patents." So a weary Alexander Graham Bell declared just two years after receiving his American rights. But while Bell felt himself growing "irritable, peevish, and disgusted with life," I have avoided a similar fate. The love, support, and intellectual companionship of Anisha Dasgupta, all of which astonish me daily, take sole credit for this. My deepest thanks go to her. I am also grateful to my two small sons, Nathaniel and Arthur, although precisely for what remains a great mystery.

Index

Aldrich, Charles, 102
American Bell Telephone Company: as monopoly, 2–3, 8, 58; incorporation of, 56–57; commitment to decentralization, 174–175; organization of, 174–175; and integration, 176, 180; ownership in individual firms, 176–177; and telephone development, 178; manufacturing arrangements, 178–179; relationship with operating companies, 179–180, 181, 201; expiration of patents, 185, 191; hostility to monopoly of, 191; prolonging of patent control, 191–195; attempt to exploit Edison patent, 192–193; inventions by employees, 194; loss of lawsuits, 194; Patent Department, 194; end of legal control over telephone, 195; dominance of, 199–200; victory over independents, 200–203; application of monopoly power, 201; reorganization of, 201
American Bell Telephone Company v. Dolbear, 68–69, 70, 79. See also *Telephone Cases*
American Bell Telephone Company v. Spencer, 67–68, 69
American Speaking Telephone Company, 52, 53
American Telephone and Telegraph Company (AT&T), 180–181, 201, 202, 208
Antitrust law, 87, 105, 108

Arc light manufacturers, 175
Aston, Theo, 135, 136, 153
Atlantic and Pacific Telegraph Company, 36
Atlantic Works v. Brady, 71
AT&T (American Telephone and Telegraph Company), 180–181, 201, 202, 208
Automobile patent, 100–101
Ayrton, William, 153

Bailey, Marcellus, 39, 40, 43, 88
Baking soda, 92
Barbed wire, 93
Bar codes, 101
Barrett, William Fletcher, 153
Barristers. *See* Patent bar, British; *individual barristers*
Bassano. See *United Telephone Co. v. Bassano & Slater*
Bate Refrigerating Company v. Sulzberger, 193
Batten, John W., 114, 122
Baxter, Judge, 97
Bell, Alexander Graham: criticism of, 2–3; reputation of, 3, 7, 82; background of, 37; research facilities for, 38; sponsors of, 38–39; as pioneer inventor, 63; overseas rights of, 110; and Telephone Company, 114–115; awarded Hughes Medal, 130; history's view of, 206. *See also* Bell patent; Patents, British

Bell patent: irregularities of application, 42–43; specifications of, 43; voice transmission in, 43, 46; second, 46–47; exploitation of, 49; Supreme Court's upholding of, 58; contentions of, 59; prominence of, 59; representation of, 65–66; fifth claim in, 66, 68, 131, 140; as pioneer patent, 68, 81; scope of, 69, 84; and claims of prior invention, 72–73; motivations for challenging, 73–74; and official misconduct, 88; expiration of, 185. *See also* Government suits; Patents, British

Bell Patent Association, 38, 46–47

Bell System, 163–164, 168, 182

Bell Telephone Annex, 96

Bell Telephone Company: formation of, 49; reliance on private finances, 52; organization of, 52–53; financial weakness of, 54–56; hostility to monopoly of, 78. *See also* American Bell Telephone Company; National Bell Telephone Company

Bell Telephone Manufacturing Company, 128

Berliner, Emile, 55, 99, 101

Berliner patent, 99–100, 101–102, 103–104, 105–106, 191–192, 195. *See also* Microphone

Berliner suit, 102–106

Blake, Francis, 55, 122. *See also* Transmitters

Blanchard, Thomas, 23, 32, 209

Blanchard v. Sprague, 28–29

Blatchford, Samuel, 79

Blodgett, Henry, 92

Boorstin, Daniel, 4

Boston, research in, 38

Bouverie, Edward Pleydell, 118, 123, 124

Boycotts, 189, 190

Bradley, Joseph, 63, 70, 82–84

Bramwell, Frederick, 137–138, 150, 153

Brand, James, 114, 124

Brewer, David, 104, 105

Britain: telephone patent struggle in, 9; patent grant in, 13–14; suspicion of grants in, 28; patent for undulatory current in, 41, 42; scire facias actions, 91; international patenting in, 109–110; Bell's plans to apply for patent in, 110–111; publicity of telephone in, 111–112, 113, 140, 146, 149; Post Office, 112–113, 122, 125–126; sale of telephone patent in, 113–114; Hubbard's work in, 115–116; Edison in, 116–124; vulnerability of Bell's patent in, 117; debut of Edison's telephone in, 118; Edison Telephone Company of London, 118–124; exploitation of Edison's patent outside London, 119–120; representation of Edison in, 120; Bell's patent in, 121, 130, 139–142, 144, 145–146, 154; competition in, 121; telephone competition in, 121; merger of telephone companies in, 122–124; telephone industry in, 127; telephone companies in, 128–129; Edison's patent in, 130, 142–144, 146–148, 150, 152–155, 156, 193; view of Bell in, 130; legal theory of telephone invention in, 130–131; patent system in, 131–135; possibility of prior publication in, 140; view of inventors in, 159. *See also* Litigation, British; Patents, British

Brown, George, 42, 110–111, 139

Business. *See* Corporations

Butler, Benjamin F., 93

Canada, telephone industry in, 127

Cancellation of patents, 86–87, 91, 99. *See also* Government suits

Capitalists, patentees' relationships with, 126

Carpenter, George, 103

Caveats: Gray's, 42, 46, 67, 88, 111; Meucci's, 72; Berliner's, 101

"Central claiming," 228n10

Centralization, impetus toward, 167

Central Union Telephone Company, 201

Chandler, Alfred, Jr., 171–172, 175, 203, 204

Chase, Salmon P., 30

Chellis, Edgar, 73–74

Chicago, 201

Choate, Rufus, 30

Cities: focus on, 187, 189; cost of service in, 189–190; Bell's dominance in, 200

Clayton Act (1914), 105

Cleveland, Grover, 90, 95

Cochrane v. Deener, 70

Collusion, among corporations, 107
Colt, LeBaron, 98
Colt, Samuel, 29
Competition: and patents, 48; in telephone industry, 51–52, 53–56, 169, 174, 195–203; in Britain, 121; failures of, 166–167; and networks, 166–167; Bell's expectations of, 196; and growth, 198; of networks, 198; report on, 203
Conferences, 181
Congress: intervention in patent matters, 29–30. *See also* Extension of patent term; Government
Conkling, Roscoe, 75
Constitution, U.S., 14
Constitutional convention, 14
Continental Telephone Company, 128
Converse, Elisha, 49
Cooke, Conrad, 150
Cooper, Carolyn, 23
Cooper, Peter, 49
Coordination, in telephone industry, 178–182
Cornish, W. R., 157
Corporations: courts' support for, 76; collusion among, 107; use of restrictive licensing, 107; in Patent Office centennial, 207; use of patents, 207–211; intellectual property strategies, 211
Corruption: and Bell patent, 76–77, 78. *See also* Scandals
Cotton gin, 16–17
Courts: and prejudices, 63–64; support for corporations, 76. *See also* Judges
Cozens-Hardy, H. H., 153
Cushman, Sylvanus, 72, 73, 191

Decentralization, 174–175, 181, 183–184
Department of Justice, Bell Telephone Annex, 96. *See also* Government suits
Dewhurst, George, 114, 115
Diaphragm, 152, 153, 155
Dickerson, Edward N., 65, 75, 96
Dickinson, Don, 74, 81
Disclaimers, 140–141, 143–144
Distribution, 172
Dolbear, 68–69, 70, 79. *See also Telephone Cases*
Dolbear, Amos, 52, 67, 69, 78, 153

Dowd, Peter, 55, 64
Dowd case, 55, 56, 64–68
Drawbaugh, Daniel, 78, 81, 99, 210
Drawbaugh case, 73–75, 79, 81, 101–102, 191. *See also Telephone Cases*
Dual service, 198

Eaton Telephone Company, 68
Edison, Thomas: work for Western Union, 36, 37, 40, 51, 52; transmitter of, 54, 117, 120, 122, 141, 152, 153, 192–193; Bell's competition with, 109; in Britain, 116–124; electromotograph, 117, 120, 122, 142; international standing of, 118; representation of in Britain, 120; telephone, 122; response to merger proposal, 123; British patents of, 130, 142–144, 146–148, 150, 152–155, 156, 193; electric lamp, 157, 193; judges' view of, 159–160; as system-builder, 168; delayed grant to, 192
Edison Telephone Company of Europe, 128
Edison Telephone Company of London, 116, 118–124, 143
Edmunds, George, 74, 75, 81
Electrical induction coil, 100
Electrical industries, 183–184
Electrical Review, 151, 155, 157, 191, 198
Electricity, research on, 38
Electromotograph, 117, 120, 122, 142
Ellsworth, Henry, 24
Employees, in patent disputes, 63
Enforcement, campaigns of, 21. *See also* Litigation
England. *See* Britain
English Mechanic, 141
Entrepreneurs, inventors as, 126
Ericsson, L. M., 128
Europe, patent laws in, 133. *See also* Britain; *individual countries*
Evans, Oliver, 17–18
Evenson, A. Edward, 4
Exchanges, 173, 174, 185
Extension of patent term, 17, 21–24, 26–27, 29, 30

Farmers, hostility of, to patents, 94
Field, Stephen, 83

Filament, 155
Filing, priority of, 39–40, 41
Firm, theories of, 183
First-mover advantage, 203, 204
Fitch, John, 15–16
Flour manufacturing, 17–18
Forbes, James Staats, 124
Forbes, William H., 53, 177–178, 190
Formalism, 158
France, telephone industry in, 127
Franchises, 51
Fraud: alleged, in Bell patent, 76, 78, 80–81; role of, in arguments before Supreme Court, 80–81; and government intervention in Bell patent, 87–88; legal logic for claiming, 88–89; and cancellation, 91; and lack of clarity in law, 91–92, 107; evidence of, 96; use of charge of, 107. *See also* Government suits; Scandals
Free trade, and opposition to patent system, 133
Fry, Justice, 153–155, 156
Functional integration, 175

Garland, Augustus, 77, 80, 89, 90, 95
Garnet, Robert, 177
General Electric Company, 193
Germany, 51, 127
Gifford, George, 32
Glidden patent, 93
Globalization, 9–10
Gold and Stock Telegraph Company, 36, 52
Goldsmid, Julian, 118
Goodyear, Charles, 25, 28, 30, 32, 48
Gould, Jay, 36, 56
Gouraud, George, 117–118, 119, 120, 123, 124
Government: and cancellation of patents, 86–87, 91; attempts to draw into patent actions, 89–91; role of, in policing patent power, 107
Government suits: and populism, 85; motives for seeking, 92–93; and politics, 93; avenues for, 94–95; dismissal of, 97, 98; choice of jurisdiction in, 97–98; Bell's defense in, 98; as channel for private interests, 106; as antimonopoly weapon, 107. *See also* Berliner suit; Telephone Cases
Gower, Frederic, 148
Granger movement, 78
Graves, Edward, 112
Gray, Elisha, 37, 52, 80; efforts to patent harmonic telegraph inventions, 39–40; patent for transmitting vocal sounds, 41–42; caveat filed by, 42, 46, 67, 88, 111
Gray, Horace, 68–69, 77, 79, 82, 238n74
Great India-Rubber Case, 26, 30
Grove, William R., 138, 160, 136

Hamilton, Walton, 203, 207
Harding, George, 32, 74
Harlan, John Marshall, 83, 238n74
Harmonic Telegraph Company, 52
Harrison, Cox-Walker & Co., 151–154
Hill, Lysander, 73, 75, 80–81
Home, Richard, 116
Hopkinson, John, 137, 153
Horsford, Eben Norton, 92
Howe, Elias, 27, 30, 32
Hubbard, Gardiner Greene: promotion of Bell, 38–39; and filing of Bell's patent, 42; and undulatory current patent, 42; offer of rights to telephone, 49; finances of, 52; position in Bell Telephone Company, 52; submission of Bell's application, 111; and Telephone Company, 114, 115–116; and Bell's overseas rights, 115; work in Britain, 115–116; organization of International Bell Telephone Company, 128; financing of exchange operations, 174
Hudson, John E., 182, 192
Hughes, David, 146–148, 150
Hunnings, Henry, 151–152, 153
Hussey, Obed, 26

Imray, John, 137, 150, 153
Indiana, 190, 191. *See also* Midwest
Industrial Revolution, 134
Industrial revolution, second, 7–8, 9–10, 208
Industry, 13–14. *See also* Corporations

Innovation: as business strategy, 36; nature of, 60; corporate organization of, 159
Intellectual property: corporations' use of, 7; relation with monopoly, 33-34. *See also* Patents
Interference, 40, 42
International Bell Telephone Company, 128
Invention: role of law in history of, 3-4; democracy of, 19; and litigation, 33; nature of, 60, 62-63; and patent law, 60-61; originality of, and patents, 61-62; relation with monopoly, 84; profitability of, 100; vs. improvement, 157, 159; commercialization of, 172, 209; by Bell employees, 194; corporate control of, 207
Invention, heroic, 63, 133-134, 159, 208
Invention, priority of, 39-40, 41, 42, 72
Invention, simultaneous, 42
Inventor: as independent operator, 37, 212; as user vs. manufacturer, 48; popular image of, 63; as entrepreneur, 126; view of, 133-134, 208; in Patent Office centennial, 207
Inventor, heroic, 63, 133-134, 159, 208

Jefferson, Thomas, 14-15, 17
Jenkin, Fleeming, 150
Jessel, George, 138, 156, 159
Johnson, Edward H., 120, 122, 123, 124
Johnson, J. H., 143
Johnson, Reverdy, 30
Jones, George, 80
Journals: *Scientific American*, 68, 70, 73; *English Mechanic*, 141; *Electrical Review*, 151, 155, 157, 191, 198; reaction to decisions, 154, 155, 157; *Western Electrician*, 194
Judges: role of, in patent cases, 76; allegations of impropriety of, 77, 79; views of patents, 104-105, 108; in Britain, 138; and formalism, 158; views of inventors, 159. *See also individual judges*

Keller, Charles M., 31
Kellogg, Milo G., 102

Lamar, L. Q. C., 95-96, 99
Lathe, 23
Lawyers: elite, in patent cases, 30-31; development of specialists, 31-32; in Drawbaugh case, 74; in government suits against Bell, 96; patent bar, British, 136-137, 138, 153, 160-161; authority of, in British patent law, 160-161. *See also individual lawyers*
Leasing plan, 50-51, 116
Lemelson, Jerome, 101
L'Enfant, Pierre, 11
Letters patent, 13-14
Licenses/licensing: incentives for, 48; restrictive, use of, 107; and integration of telephone industry, 171; and exchange format, 174; permanent, 174-175; restrictions on, 178
Lighting enterprises, 175-176
Lincoln, Abraham, 7, 31, 217n24
Linguistics, in British litigation, 158
Litigation, American: and term extension, 22-24, 26-27; and reissue, 25-26; consequences of campaigns of, 27-32; and inventions, 33; Bell's second patent in, 47; interpretation of patents in, 47-48; organized response to, 92-93, 195; and power of corporations, 209-210; and questions about monopoly, 210; decline in, 211; comeback of, 211-212. *See also* Government suits; *Telephone Cases*; *individual cases*
Litigation, British: number of patent cases, 135-136; expert witnesses, 137, 138, 150, 152-153, 154, 155, 160, 161; judges, 138; by Edison Telephone Company, 143; *United Telephone Co. v. Harrison, Cox-Walker & Co.*, 148, 151-155, 156; *United Telephone Co. v. Maclean*, 148, 149-151, 152; desire for Scottish decision, 149-150; reactions to decisions in, 154, 155-156, 157; *United Telephone Co. v. Bassano & Slater*, 154-155; favor for pioneer grants, 156-160; *Proctor v. Bennis*, 157; imbalances in, 160-161
Lobbying, and term extensions, 29
Lockwood, Thomas, 194
London and Globe Telephone Company, 151

Long-distance service, 180–181, 187
Lowell, John, 68, 70, 79
Lowell Syndicate, 176
Lowrey, Grosvenor, 81, 96–97, 99, 195–196
Lubbock, John, 118
Lynd, Robert, 207

MacDougall, Robert, 199
Maclean. See *United Telephone Co. v. Maclean*
Maclean, Alexander, 149
Macroinventions, 60
Madden, Oscar, 53
Madison, James, 14, 17
Magneto telephone, 46, 54, 139, 141
Market entry: and patent control, 171; control of, 182; threat of, 190; by independent telephone companies, 195; barriers to, 204
Marketing, 172, 174
Mason, Charles, 31
Matthews, T. Stanley, 79
McCormick, Cyrus Hall, 26–27, 30, 31, 32, 172
McLaren, Lord, 150–151
McLean, John, 23
"Men of Progress" (Schussele), 32
Mergers, 176
Metallic circuit, 180, 187
Meucci, Antonio, 72, 73, 191, 206
Microinventions, 60
Microphone: Drawbaugh's claim to, 101; use of term, 146, 148. *See also* Berliner patent; Variable resistance
Midwest, 190, 191, 196–197, 199, 201
Miller, Phineas, 16–17
Miller, Samuel, 91, 98–99
Milling, 17–18
Molecular Telephone Company, 78. See also *Telephone Cases*
Monarchy, and patents, 13
Monopoly: American Bell as, 2–3, 8, 58; relation with intellectual property, 33–34; over telephone, 49; hostility to, 78, 94; relation with invention, 84; justification of, 168; explanation of, 168–169; as inevitable, 168–169; failure to sustain, 203. *See also* Antitrust law; Scandals

Monopoly, natural, 166, 167, 203
Monopoly, patent: attempts to achieve, 22–27; consideration of, in issuing of patents, 103–104; nature of, 107; acceptability of, 108; and system, 177–181; expiration of, 191; and lasting market power, 203–204; dependence on existing power, 211
Monopoly, royal, 13
Morgan-Brown, William, 139
Morris, John, 114
Morse, Samuel F. B., 24, 30, 32
Moulton, John Fletcher, 136, 137, 153
Mowry v. Whitney, 91, 93, 99
Mugwumps, 94

National Bell Telephone Company, 53, 56. *See also* American Bell Telephone Company
National Improved Telephone Company, 89, 95–96
National Telephone Exchange Association (NTEA), 181, 191
Nelson, Thomas, 98
Network effect, 166
Networks: concept of, 165; and monopoly, 166, 170; and competition, 166–167, 198, 201–202; and public policy, 167; in history of telephone industry, 168–169; contribution of patents to, 170; development of, 187; underground, 187
New England Telephone Company, 53
New England Telephone & Telegraph Company, 176
Newspapers, 79–80, 155–156. *See also individual newspapers*
New York City, 200
New York City Board of Trade, 78
New York Herald, 79
New York Times, 80, 192, 193
Nonobviousness, 71
North, Justice, 154, 155, 160

Oil, 92–93
Oligopolies, corporate, 211
O'Reilly v. Morse, 25, 29, 69–70

Organization, 172, 175
Oriental Telephone Company, 128
Orton, William, 36, 40, 49, 51
Otto gas engine, 156–157
Overland Telephone Company, 78. See also *Telephone Cases*

Page, Charles, 100
Pan-Electric Telephone Company, 77, 80, 89–91, 92, 95–96, 99. See also Government suits
Parker, Austin, 22, 28, 30
Parker, Zebulon, 22, 28, 30
Patent, Bell's. See Bell patent
Patent, for telephone. See Bell patent; Patents, British
Patent, narrow, 62
Patent, pioneer: concept of, 62; Bell's patent as, 68, 81; decline in reliance on, 107; judicial favor for, 156–160; reliance on, 210
Patent, submarine/sleeping, 100–101
Patentability, limitations of, 69–70
Patent Act (1790), 14–15, 17–18
Patent Act (1793), 15, 16, 18, 19–20
Patent Act (1836), 20, 61
Patent Act (1870), 100
Patent agents, in London, 110
Patent bar, British, 136–137, 138, 153, 160–161. See also *individual barristers*
Patent claims, interpretation of, 47, 61, 72, 131, 156–158. See also "Central claiming"; Patent, pioneer; "Peripheral claiming"
Patent control: and creation of monopoly, 170; effects of, 171; and market structure, 171; and power over interdependent technology, 182–183; and development of telephone industry, 183–184; Bell's prolonging of, 191–195
Patent defense associations, 92, 195
Patent law, American: hundredth anniversary of, 1–2; role of, in American history, 4; omission from history, 5–6; reputation of, 6; significance of, in history, 6–7; origins of, 13; *Scientific American* on, 70; doctrinal shifts in, 71; possibility of repeal of, 94; lack of, 128;

foreign expiry provision, 193. See also Patent system, American
Patent law, British, 157–158, 160–161. See also Patent system, British
Patent Office (British), 132
Patent Office (U.S.), 11–12, 20, 205, 207
Patent pools, 108, 211
Patent professionals, 21, 110. See also Lawyers
Patent rights, 104–105, 108; and national legal systems, 10; exchange of, for stock, 173, 175, 177; influence of, 209
Patents, American: and rise of big business, 8; geographic distribution of applicants, 18–19; economic factors in, 19; quality of, 19; position in American life, 20–21; growth in, 21, 32–33, 62; scope of, 25, 61; judicial views of, 28–29; malleability of, 47–48; practical implications of, 48; exploitation of rights of, 48–49; Brewer's view of, 104; issued to Britons, 109; as tool of business, 206–207
Patents, British: Bell's, 121, 130, 139–142, 144, 145–146, 154; Edison's, 130, 142–144, 146–148, 150, 152–155, 156, 193; numbering of, 239n6
Patent system, American: beneficiaries of, 12; visibility of, 12; early, 13–18; development of, 21; dilemma of, 83–84; private nature of, 98; strength of, 107–108
Patent system, British, 131–135
Pennsylvania, 191
People's Telephone Company, 74, 80
"Peripheral claiming," 228n10
Perkins, John McClay, 79
Politics: role of, in scandals, 80; and *Telephone Cases* decision, 83; and government suits, 93, 106. See also Congress; Government
Pollok, Anthony, 39, 40, 42
Post Office (Britain), 112–113, 122, 125–126
Precedent, focus on, 158
Preece, William, 112, 113, 126, 147
Price, J. Lowell, 118
Price-fixing, 108
Principles, in patents, 69–70
Proctor v. Bennis, 157
Proprietary groups, 173–174
Publication, prior, 140, 145–146, 153

Publicity: and Morse's bid for patent monopoly, 24–25; of telephone, in Britain, 111–112, 113, 140, 146, 149

Quilter, W. Cuthbert, 114

Railroads, 48, 51, 78, 94
Reaper, 26–27, 30, 31, 172
Receivers, 46–47, 117, 120, 122, 142, 144
Regulation, and Bell's dominance, 203
Reis, Philipp, 67, 68, 73, 81, 96–97, 130, 145, 150, 152, 153, 206
Reissue, 21, 25–26, 47, 71, 140–141
Research, industrial, 37–38, 208, 211
Revocation, 86–87, 91, 99. *See also* Government suits
Revolver, 29
Reynolds, William H., 111, 112, 113, 114, 126
Roberts, E. A. L., 92–93
Royal Society, 118, 130
Rubber, vulcanized, 25, 28, 30
Rumsey, James, 15–16
Rural areas, telephone service in, 197
Rural economies, 94

Sales, of new inventions, 172
Sanders, Thomas, 38, 52
Scandals, 79–80, 95. *See also* Fraud
Schussele, Christian, 32
Scientific American, 68, 70, 73
Scire facias actions, 91
Scotland, 149–150. *See also* Britain
Scott, Adam, 114–115, 116
Scribner, Charles E., 194
Selden, George, 100
Seward, William H., 30, 31–32
Sewing machines, 172
Shaw, George Bernard, 116–117, 122
Shoe machinery, 50
Shulman, Seth, 4
Siemens, Werner von, 127
Skowronek, Stephen, 106
Smith, Chauncey, 50, 65, 66–67
Speech transmission. *See* Voice transmission

Spencer, Albert, 68
Stanton, Edwin, 31
Steamboat technology, 15–16
Stock, 173, 175, 177
Storrow, James J.: on telephone leasing plan, 50; description of, 65; in Drawbaugh case, 75; in *Telephone Cases*, 81; on intellectual property, 98; on telephone industry, 162; death of, 192; on prolonging patent control, 192
Strong, James, 24
Sun (New York), 80
Supreme Court: and scope of patent claims, 29; upholding of telephone patent, 58; hearing of *Telephone Cases*, 78–84; and Berliner suit, 104
Sweden, 127–128
Switchboards: debut of, 165; improvements in, 186; patents for, 194
Systems: Bell System, 163–164, 168, 182; concept of, 165–166; in history of telephone industry, 167–169; idea of, 168; and monopoly, 168, 170, 177–181; contribution of patents to, 170

Taylor, Robert S., 102–103, 104
Technology: elevation of, in Britain, 134; social construction of, 169; corporate control of, 207; social organization of, 207
Telegraph: Morse's bid for patent monopoly on, 24–25, 30; Atlantic and Pacific Telegraph Company, 36; Gold and Stock Telegraph Company, 36; and patents, 36; innovation in, 36–37; competition in, 56; in Britain, 112. *See also* Western Union
Telegraph, harmonic, 39–40
Telephone: inventor of, 4–5; overseas market for, 9; monopoly over, 49; proprietary interest in, 49; rights to, 49; as proprietary device, 50; number of, 185; improvements in, 186
Telephone Cases: importance of, 58–59; lawyers in, 65; *American Bell v. Dolbear*, 68–69, 70, 79; and changing laws, 69–70; Drawbaugh case, 73–75, 79, 81, 101–102, 191; scale of, 74; decisions in, 75; Supreme

Court's hearing of, 78–84; and Bell's reputation, 82; opinions in, 82–84; use of term, 227n1. *See also* Bell patent
Telephone companies: models of, 74; injunctions against, 75; support for, 78; growth of, 105; in Europe, 128; in Britain, 128–129; and patent rights, 183; business strategy, 186; profits of, 186; development of networks, 187; priorities of, 187, 188–189; operating without Bell permission, 190–191; market entry by, 195; markets of, 196; approach to service, 197; numbers employed by, 199; cooperation with Bell, 202; purchase of by AT&T, 202. *See also individual companies*
Telephone Company, Ltd., 113–116, 121, 122–123
Telephone equipment: manufacture of, 49, 178–179; quality of, 54–55; manufacture of, in Europe, 128; change in, 197–198. *See also* Western Electric Manufacturing Company
The Telephone Gambit (Shulman), 4
Telephone industry: structure of, 50–51, 162–163; in Europe, 51, 127–128; competition in, 51–52, 53–56, 169, 174, 195–203; Bell's conception of, 164–165; economic environment of, 166–167; historians on form of, 169–170; integration of, 171; local nature of, 172–173; patent-based relationships in, 173–177; decentralization in, 174–175, 181; consolidation movement, 176; mergers in, 176–177; ties between firms, 177–182; coordination in, 178–182; growth of, 198–199; marketing developments, 199; barriers to entry, 204. *See also* American Bell Telephone Company; Networks; Systems
Telephone lines, 187. *See also* Networks
The Telephone Patent Conspiracy of 1876 (Evenson), 4
Telephone Protective Association, 195
Telephone rates: regulation of, 78, 190; in Britain, 122; complaints about, 189, 190; of independent companies, 197
Telephone service: promotion of, 49; franchises, 50, 51; Western Union's, 51–52; relationship with patents, 127–129; tendency toward monopoly, 166; as

indivisible, 168; organizational model of, 172; long-distance service, 180–181, 187; market for, 186–187; complaints about, 189; expansion of, 196; by independent companies, 197; dual service, 198
Thompson, Silvanus P., 153, 155, 156
Thomson, William, 111, 113, 140, 146, 153
Thornton, William, 15, 19
Tilghman, Richard, 70
Tilghman v. Proctor, 70
Torpedo, nitroglycerin, 92–93
Townsend, William, 62
Transfers, international, 127–129, 144
Transmitters: Edison's, 54, 117, 120, 122, 141, 152, 153, 192–193; Blake's, 55, 122, 141, 247n51; control of, 99–100; and Edison's British patent, 117, 142, 144, 146–148; and British patents, 141–142; in British litigation, 151–152, 153; Hunnings's, 151–152, 153
Tribune (New York), 80
Tying requirements, 105
Tympan, 152, 153, 155
Tyndall, John, 118

Undulatory current, 40–41, 43, 64, 130–131, 141, 142
United States v. American Bell. *See* Government suits; *Telephone Cases*
United States v. Colgate, 92
United Telephone Company Ltd., 124, 131, 144, 145, 148–156
United Telephone Co. v. Bassano & Slater, 154–155
United Telephone Co. v. Harrison, Cox-Walker & Co., 148, 151–155, 156
United Telephone Co. v. Maclean, 148, 149–151, 152
Utility companies, 173, 175

Vail, Theodore, 53, 180–181
Van Ness, William, 19–20
Variable resistance: as principle of transmitter, 55, 141–142, 150; principle alleged stolen by Bell, 76, 88, 96; as basis of Edison's British patent, 142–143, 146, 151–152. *See also* Transmitters

The Visible Hand (Chandler), 171–172
Voice transmission: development of, 35; Bell's research on, 40–41; in patent for telephone, 43, 46; breakthrough in, 46; Bell's monopoly on, 69

Waite, Morrison R., 58, 82, 99
Wallace, John, 91, 92
Wallace, William, 75, 81
Washburn and Moen, 93
Washington, D.C., plan for, 11
Waterhouse, Alfred, 118
Waterwheel, 22, 28, 30
Watson, Thomas, 40, 42, 153, 178, 194
Watt, James, 2, 82, 134, 159, 209
Webster, Daniel, 30
Webster, Richard, 136–137, 143, 153, 156
West, hostility to patents in, 94
Western Electrician, 194
Western Electric Manufacturing Company, 102, 128, 178, 194
Western Union: dominance of, 35–36; rights to harmonic telegraph, 37; and competition, 39; rejection of telephone rights, 49, 125–126; telephone services, 51–52; patents of, 54; competition with Bell Company, 54–56; withdrawal from telephone field, 55–56, 67; transfer of telephone patents to Bell, 56; in Dowd case, 67; and development of telephone industry, 125–126; and Stearns patent, 230n53
White, Arnold, 122, 123
Whitman, Charles S., 99
Whitney, Eli, 16–17
Wilber, Zenas Fisk, 43, 88, 96
Williams, Charles, 38, 178
Wilson, James G., 24, 31
Winans v. Denmead, 29
Witnesses, expert, 137, 138, 150, 152–153, 154, 155, 160, 161
Woodbury patent, 92, 93, 100
Woodcroft, Bennet, 133
Wood-planing machine, 23–24, 28, 30, 31, 92, 100
Woodworth, William, 23–24, 28, 30, 31
Working requirement, 86–87
World (New York), 80